Studies in Applied Philosophy, Epistemology and Rational Ethics

Volume 29

About this Series

Studies in Applied Philosophy, Epistemology and Rational Ethics (SAPERE) publishes new developments and advances in all the fields of philosophy, epistemology, and ethics, bringing them together with a cluster of scientific disciplines and technological outcomes: from computer science to life sciences, from economics, law, and education to engineering, logic, and mathematics, from medicine to physics, human sciences, and politics. It aims at covering all the challenging philosophical and ethical themes of contemporary society, making them appropriately applicable to contemporary theoretical, methodological, and practical problems, impasses, controversies, and conflicts. The series includes monographs, lecture notes, selected contributions from specialized conferences and workshops as well as selected Ph.D. theses.

Advisory Board

More information about this series at http://www.springer.com/series/10087

Gianluca Caterina · Rocco Gangle

Iconicity and Abduction

Springer

Gianluca Caterina
Department of Mathematics
Endicott College
Beverly, MA
USA

Rocco Gangle
Department of Humanities and Philosophy
Endicott College
Beverly, MA
USA

ISSN 2192-6255 ISSN 2192-6263 (electronic)
Studies in Applied Philosophy, Epistemology and Rational Ethics
ISBN 978-3-319-44244-0 ISBN 978-3-319-44245-7 (eBook)
DOI 10.1007/978-3-319-44245-7

Library of Congress Control Number: 2016948109

Printed on acid-free paper

This Springer imprint is published by Springer Nature
The registered company is Springer International Publishing AG Switzerland

Contents

Introduction

The present book aims to examine the role of iconic signs in scientific hypothesis formation and to show how the relationship between the semiotics of iconicity and the logic of abductive inference finds a natural mathematical setting in category theory and, more specifically, in the categorical theory of topoi and the nonclassical logic of sheaves.

The book's thesis is that the core method of category theory, which lifts properties characterizing individual objects to structural properties characterizing systems of relations linking individuals to one another, helps to illuminate the creative, context-dependent and tentative nature of abductive inference. In particular, category theory sheds new light on how and why mathematics itself is so often successfully employed in scientific hypotheses and their experimental testing.

A helpful touchstone for indicating the parameters and stakes of the present book is Peter Galison's *Image and Logic*, a text that has become something of a minor classic in contemporary philosophy of science. Galison describes his book as being about "the machines of physics" (xvii). Namely, it concerns the complex interplay of social, theoretical and engineering issues in the twentieth-century development of experimental and measurement apparatuses for microphysics, from cloud chambers to supercolliders. Structuring this multilayered genealogy of modern atomic and subatomic physics is the duality of "image" and "logic" indicated by the book's title, the divergence of the evidential criteria in play for the field of physics as a whole between on the one hand photographic images carrying a maximum of relational and relatively unformatted data requiring selective interpretation by observers and on the other hand digitized data such as that provided by particle counters immediately subject to purely logical and mathematical analysis.

What is ultimately at stake for Galison in studying the productive tension between image-based and logic-based data is the theorization of the concrete processes whereby scientific knowledge and the objects of scientific study are co-constructed via hypothesis and experimental testing in local environments (laboratories, seminars, engineering workstations) that must function more or less successfully in their own right as well as globally coordinate their successes appropriately with one another. Galison's is an inquiry into the veritable conditions of possibility for collective scientific endeavor at the kinds of institutional and financial scales required by contemporary physics, where scientific epistemology is necessarily entwined with political, social, economic and technological concerns. This far-ranging research project has been continued and extended in other works by Galison, notably the collaboration [1] with Lorraine Daston.

In an analogy with Galison's "machines of physics", the present book may be said to concern itself with the "machines of abduction" in scientific inquiry and theorization, particularly its widely prevalent and increasingly sophisticated *mathematical* "machines". Whereas Galison focuses on the *detector* as a type of material apparatus that in twentieth century physics played the role of epistemological bridge "between the microworld and the world of knowledge" (xviii), we examine the Peircean notion of *iconicity* as a semiotic structure linking conceptual fields organized by logical relations on the one hand and real domains organized by structural and causal ones on the other such that the type of linkage itself thereby provides important clues as to the possibility and internal functioning of abductive inference within processes of scientific theory construction. In particular, we aim to show that the semiotics of iconicity as realized in the logic of presheaves and sheaves and rigorously formulated in the language of categorical topoi goes quite some distance in explaining the necessity and the success of the employment of mathematical tools in the scientific investigation of nature.

Like Galison, our primary concern is with the processes whereby scientific knowledge is actually generated. Yet while Galison focuses on the particular field of subatomic physics, we wish to cast a somewhat wider net that hopes to capture essential features of scientific reasoning about the natural world as such. We hope that the apparently unreasonable ambition signaled by this much larger scope of inquiry may perhaps be justified by the fact that we look almost exclusively to the unique role played by mathematics in abductive reasoning in science. We are interested in how the most concrete and situated aspects of scientific hypothesis formation might be modeled in and better understood through techniques and constructions immanent to the abstract world of contemporary mathematics. Iconicity functions as both method and object of investigation here. We subscribe equally to each of the triad of slogans "Mathematical abstraction entails maximum cognitive generality"; "Scientific reasoning concerns what is real and is always itself part of a concrete natural process"; and "Rigorous science requires mathematics".

The overall argument consists of the coherence of three claims: (1) that there is an essential relationship between abductive reasoning and iconicity; (2) that this relationship helps to explain how abductive inferences work; and (3) that category theory and, more particularly, topos theory offers a rich and natural formal domain (a type of mathematical as well as logical "language") for exploring this connection between iconicity and abduction in a rigorous yet also intrinsically abductive way. We take each of these three claims in turn. We claim first of all that there is an important, even essential, relationship between abductive reasoning and iconicity (the semiotic mode whereby one thing signifies or represents something else via a common relational structure); second, that this relationship sheds light on how abduction works, and in particular helps to explain why abductive inference has remained so resistant to formalization; and finally that category theory, and especially the theory of topoi within it, provides a formal representational environment (a "language") with particular utility for exploring this connection between iconicity and abduction in a mathematically tractable way.

We proceed by examining two philosophers who engage the theoretical problem of the truth/knowledge relation from an ontological standpoint and who make extensive use of mathematical models and techniques: Charles Peirce and Alain Badiou. In their very different work, we identify a common pattern: the difference between Boolean and non-Boolean Heyting algebras and the use of the "geometrical" internalization of logic as a way to expand or modalize a formal situation from within.

Both philosophers use formal systems in an onto-epistemic register. In other words, both of them employ formal languages with the explicit intention of engaging real world processes of reasoning and also of tracking the problem of non-deductive inferences and creative hypotheses. In Peirce, this appears as the coordination of his general theory of semiotics and particularly his notion of diagrammatic mathematics with the iconic logical system of Existential Graphs. In Badiou, this appears as the use of Zermelo–Fraenkel set theory and Paul Cohen's forcing technique to model ontology and abductive inference in *Being and Event*, and his turn in *Logics of Worlds* to the use of sheaves and—in a restricted register—topoi to model a logic of worldly appearance, or "phenomenology". We aim to coordinate these otherwise disparate philosophical approaches within the background theory of mathematical categories.

This book consolidates the co-authors' collaborative work over the past half-dozen or so years. In that time, we have been extremely fortunate to encounter researchers working in related areas whose input, criticisms and suggestions have been invaluable. We should thank first of all our brilliant collaborator at the Center for Diagrammatic and Computational Philosophy, Fernando Tohmé. We thank also Joshua Ramey and Indradeep Ghosh for inviting us to give a workshop at Haverford College in 2012. Gratitude is due as well to the circle of interdisciplinary scholars brought together at the series of Model-Based Reasoning conferences, including especially Woosuk Park, Lorenzo Magnani, and our co-panelists at the Model-Based Reasoning conference at the University of Campinas, Brazil in 2009: Ahti Pietarinen, Jaakko Hintikka and Priscila Borges. We also thank our colleagues at Endicott College, particularly Ellis Cooper, and the college itself for multiple forms of research support in the completing of this project, including one-semester sabbaticals for each of us. Finally, we wish to recognize the helpful comments and suggestions of two anonymous reviewers and the work of the editorial team at Springer/SAPERE.

A synopsis of the content and organization of each of the six chapters may aid the reader in keeping the course of the overall argument in mind while passing first-hand through its details and also to facilitate, if desired, a more pick-and-choose method with respect to specific issues and topics.

Chapter 1 examines the concept of abduction. It traces some of the contemporary lines of investigation into this notion and frames the issue for the present context. We align our approach broadly with the project of naturalized epistemology, that is, treating epistemic agents as real actors in the world subject to all manner of natural conditions. In other words, our view of knowledge is a pragmatic, experimental and fallibilist one. Within this context, however, we are especially interested in scientific knowledge and its prevalent use of mathematics. What kind of experimental activity is the scientific employment of mathematics in the investigation of nature? Because of our focus on this question, the problem of the formalization of abduction becomes particularly acute. We argue that the concept and method of iconicity—a concept that we will try to articulate mathematically through category theory and finally within topoi—provides a way to move forward constructively in addressing the formalization of abductive cognition. We examine several recent formal models to gain some bearings and sketch out a first informal model of the image of iconic epistemology as a functor-like duality, a contravariance, between structured spaces of knowledge and reality. This is the informal intuition that will motivate the mathematical modeling of diagrams as presheaves in the second chapter. Thus, the constructions that follow through the ensuing chapters are motivated essentially by this problem of formalizing abduction as iconic, and its deep intrication of logic, mathematics and epistemology.

Chapter 2 examines Peirce's concept of iconic signs, situating this key concept in the broader framework of Peirce's general theory of semiotics. The chapter culminates in a theory of diagrams as presheaves which serves as the theoretical basis for the ensuing analysis of Peirce's Existential Graph (*EG*) notation.

Chapter 3 shows how Peirce's diagrammatic logical notation of *EG* makes use of iconicity, and how some important features of this iconicity may be represented within category theory. In particular, it is shown how the *alpha* level of the system of *EG* may be represented as a category and how logical conjunction and disjunction correspond to products and coproducts respectively in this category, while material implication may be modeled by categorical exponentiation. This latter fact allows the basic "shape" of abductive inference in classical propositional logic to be characterized by a pair of adjoint functors.

Chapter 4 turns to the work of Alain Badiou and traces his model of truth-procedures as set-theoretical forcing in the manner of Paul Cohen. We interpret this model, and especially its use of generic sets in terms of the problematic of abduction and iconicity in a way that allows for the coordination of Badiou's setting of philosophical ontology in the framework of set theory with Peirce's distinct levels of Existential Graphs.

Chapter 5 examines Badiou's more recent work, *Logics of Worlds*. Here, we specify the notion of sheaves over a Heyting algebra that Badiou uses to model the "phenomenology" of logical relations in and through which the various "objects" of a world are structured together as a unified system. We clarify how Badiou's results may be formulated within topos theory and how this helps to understand the relationship between *Logics of Worlds* and the earlier use of the set-theoretical method of forcing in *Being and Event*.

Finally, Chap. 6 gathers the threads from the previous chapters by showing how topoi provide a natural mathematical setting for the nonstandard "logic" of quantum mechanics. Drawing on the work of Cecilia Flori, Michael Epperson and Elias Zafiris and David Ellerman we show how the formal expression of quantum mechanics in category theory and more specifically sheaf theory and topoi provides a basis for revising—abductively—some of the most fundamental assumptions of scientific realism. This important and far-ranging example drawn from contemporary physics illustrates how the same epistemically abductive structures represented mathematically by presheaves, sheaves and topoi are instantiated also in physical (quantum) systems, thereby opening the path towards a more naturalist interpretation of the abductive models developed in Peirce and Badiou.

A pair of appendices offer brief reminders (or, if necessary, lean introductions) for the reader of the essential elements of category theory and topos theory. Parts of chapters two, three and four have appeared previously as articles in *Synthese, Logic Journal of IGPL* and *Studies in Computational Intelligence*. For all of their gracious support, this book is dedicated to our wives and sons.

Reference

L. Daston, P. Galison, *Objectivity* (Zone Books, Brooklyn, 2007)

Chapter 1
Modeling Abductive Reasoning

1.1 Introduction

The following chapter introduces the problem of abduction understood as a form of practical reasoning. It is necessary to state up front that the scope and stakes of this problem vastly exceed what may be adequately measured or encompassed here. The ultimate concern of the present book is not with surveying the generality of abductive inference in an external manner but rather with constructing a robust mathematical framework within category theory that can help advance ongoing attempts at formal treatments of this problem itself, especially with respect to what appears to us to be the core issue at stake, that of the possibility of modeling internally to mathematics the processes within natural scientific inquiry of formal and mathematical abductive inferences as such.

With this end in view, the ensuing sections of the present chapter begin by framing the problem of abductive reasoning in terms of a pair of diagrammatic toy models of the relationship between epistemic states (consisting of various degrees of ignorance and knowledge) and potential states of affairs (consisting of various degrees of determination and indetermination). The questions of if and how the two given models might cohere with one another set the stage for the later casting of abductive reasoning in categorical and sheaf-theoretical terms. Further sections then examine some attempts to schematize the logic of abduction, from Peirce's pioneering work to the more recent studies of Gabbay and Woods, Aliseda and Magnani, and focus upon the context-bound nature of abductive reasoning and some of the stubborn obstacles involved in any attempt to formalize abductive logic. In place of such a formal schema, we present instead a sketch of what we propose to call the generic shape of abductive reasoning, and we look in particular at how that shape may be recognized in the scientific use of mathematics. Finally, the central concept of iconicity that will be developed further through the mathematical constructions of presheaves and sheaves is introduced in the chapter's final section as a guiding thread for subsequent chapters.

G. Caterina and R. Gangle, *Iconicity and Abduction*,
Studies in Applied Philosophy, Epistemology and Rational Ethics 29,
DOI 10.1007/978-3-319-44245-7_1

Readers interested in general introductions to the characteristic features of abductive inference and more detailed histories of its study should consult [1–6].

1.2 Framing the Problem

The general context for addressing the problem of abduction is the problem of the relation between knowledge and reality, that is, the fundamental problem of epistemology. From long philosophical reflection on this problem, an extremely diverse and complex space of interconnected concerns and definite positions has emerged and continues to expand and reconfigure itself. Increasingly specialized subproblems determine a variety of research programs. Rather than attempt to wade into these deep waters and to try to sketch out, before even beginning to examine our real concern, the contemporary terrain of scientific and philosophical conceptions of knowledge, much less to take sides in the various debates, it will suffice here to specify precisely which aspects of the rich fabric of epistemology are relevant to the present endeavor. In effect, what matter are just two things: (1) ordered transitions among states of knowledge; and (2) co-ordinations between states of knowledge and degrees of determination of reality. Both (1) and (2) are ordered systems, and clearly system (2) depends partially on system (1) since (2) includes (1) as one of its constituent and determinative parts. Consider the relationship between the three progressively determined knowledge states as pictured in the base of the following figure and the set of possible models consistent with those states linked to each state as pictured in the ovals above each state.

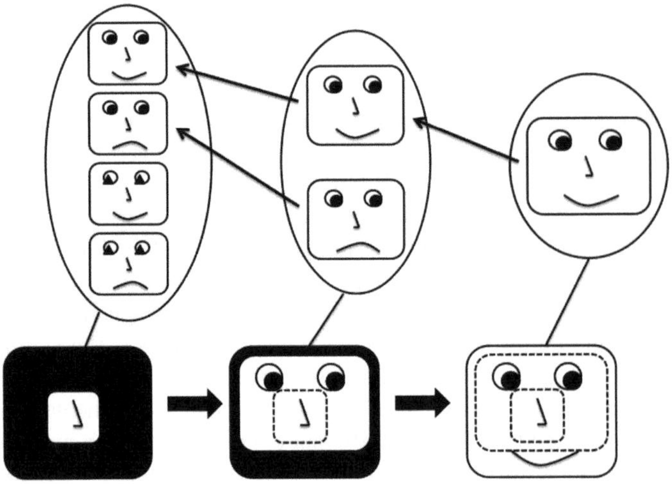

The sequence of states pictured at the base represents the progressive extension of knowledge in this particular instance. Relative ignorance, represented by the black area at each stage, shrinks as truth becomes increasingly unveiled and knowledge thereby more fully determined for the epistemic agent represented. As knowledge increases at each step, the "space" of possible differentiated states of affairs compatible with that stage of knowledge shrinks or collapses.

The simplest way to formalize (1) above in accord with this intuitive picture would be to represent it as a discrete linear order. Working with the idealization of a single epistemic agent or knowledge base monotonically increasing in a finite sequence of discrete steps, the temporal development of the system may then be represented by a finite set K of knowledge-states $k, l, m, \ldots$ where each of $k, l, m, \ldots$ is the set of what is known at that state and a linear order $<$ is defined on those elements of K. An element k of K representing a particular knowledge-state is "$<$" another such state l element of K if and only if the knowledge known at state l extends the knowledge known at k, that is, if $k \subset l$. Intuitively such a pair $(K, <)$ represents a finite "path" of knowledge, never forgetful of past knowledge (conservative) and always advancing to encompass more and more knowledge (progressive).

$$\begin{array}{ccccc} M_{K_n} & \xleftarrow{res_{K_n}} & M_{K_{n+1}} & \xleftarrow{res_{K_{n+1}}} & M_{K_{n+2}} \\ \uparrow & & \uparrow & & \uparrow \\ K_n & \xrightarrow{\subset} & K_{n+1} & \xrightarrow{\subset} & K_{n+2} \end{array}$$

Here res_{K_n} indicates the map that sends a function $f \in M_{K_{n+1}}$ into its restriction to K_n.

A space of multiple such possible paths may be constructed by relaxing the linear order $<$ on K and thereby generalizing to a partial order $\leq$. Any partial order of this kind defined for some fixed set of knowledge-states K can be embedded in the subset lattice of the power set of K.

To be sure, such models remain idealizations. Actual knowledge-agents in the world make mistakes and forget things. To enrich the model, several options readily present themselves:

1. The sequence in the base could be exfoliated and enriched. Locally monotonic sequences could be embedded in a more flexible global framework allowing for occasional nonmonotonic (and epistemically salient) discontinuities or "leaps".
2. *Degrees* of knowledge and ignorance could be built in to the model. These might reflect subjective certainty and uncertainty, forms of objective justification, or both. Bayesian techniques, for instance, might be introduced.
3. Different *types* of knowledge could be accounted for. At the very least, the distinctions necessary for fundamental epistemic dualities such as those of types and tokens or laws and cases could be introduced.
4. The models over each state could be brought to reflect the real variability of potential compatibility with knowledge. In particular, the set of numerically distinct individual models over each state should, if possible, become a more highly structured space of relations among the models.

Let us turn to an alternate conceptual framework. Using exactly the same toy construction (sets of possible models over ordered elements of a base), a slightly different perspective on knowledge may be suggested. This second perspective conceives of states of knowledge as coordinated wholes that are built up from their component parts via logically regimented relations. In the following diagram, the same space of possibilities is represented in an alternate way by treating the base structure as a "generic" or purely formal figure that represents only the way that the appropriately known components of the form (the chosen parts of the face) are coordinated into a whole. Then above each progressively determined larger part, the set of possible models for that part is determined in such a way that every such model may be suitably restricted to each and all of its constituent components.

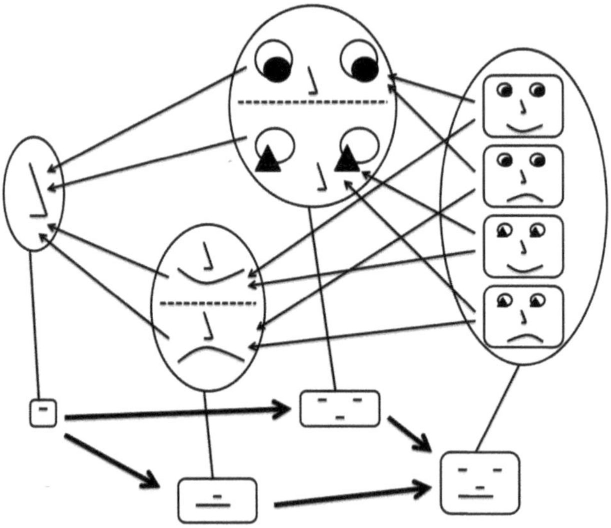

Special issues that arise here include the following:

1. How are the fragments in the base derived and how are they ordered? Simple composition of parts appears too crude in many cases.
2. Just as for the previous model, the conditions for distinguishing elements in the sets above each stage should, if possible, be built into the model as a whole.

In addition to the questions that arise for each model independently, there is the further set of questions of how the two models might productively interact. Nonetheless, taken together, all of the considerations just noted suggest a general quasi-formal problem of how to take this kind of model involving coordinated systems of composition and order and embed it in a richer space of relations in such a way that the original correlations *tend* to be conserved, but might under appropriate conditions be somewhat revised. We claim that any adequate model of abductive reasoning must take into account an analogous concern with embeddings and conservative revisions of belief-systems.

We claim furthermore that the mathematics of category theory and in particular the logic of presheaves and sheaves as expressed naturally in topoi provides a suitable formal environment for producing and testing models of explanatory hypothesis formation (abduction) that integrates both of the intuitive pictures given above as well as the relations between them. This claim and its full justification will occupy the book as a whole. In the remainder of the present chapter, we examine the concept of abductive inference in greater detail.

1.3 Defining Abduction

> It is our ignorance of things that causes all our admiration, and chiefly excites our passions.
>
> Edmund Burke, *Philosophical Enquiry into the Origin of Our Ideas of the Sublime and the Beautiful*

"What was that?" a frightened human asks upon hearing a strange sound in the limitless dark outside some well-lit interior, the inhabited space of a house or a tent or a frescoed cave. It is natural that the potent emotional cocktail of surprise and ignorance should eventuate in guessing, in positing some plausible account of what, if true *and* known, would cancel out the emotional affects of surprise and fear and render these in retrospect merely contingent effects of relative ignorance. In this way, "hypothesis-formation" in a broad and informal sense appears to stand at the root of thoughtful inquiry per se. In Vico, for instance, such a narrative of the genesis of meaning from animal fear and the seeking for explanation is itself meant to be explanatory of human history and culture. The hypothesis of hypothesis-formation itself thereby provides a plausible explanation of how rationality might have emerged from pre-rational impulses and affectivity. The notion of abduction is thus itself a kind of hypothesis.

At a first approach, abduction may be contrasted on the one hand with deduction. Deductive inference draws a conclusion from some set of given premises such that if all of the premises are true, the conclusion follows necessarily. In this sense, the conclusion may be understood to be present in the premises themselves, although only in a latent or implicit fashion. In order to "draw" the conclusion from the premises (and the implicit metaphor of "pulling" something out of some material, extracting it smoothly, is particularly apt) one typically has recourse to formal rules that present uncontroversial (within some given context of discourse!) logical truths. Thus the essential movement of deduction is one which abstracts a kind of kernel of formal relationships that hold already among the premises and then operates on this kernel (permuting, condensing, elaborating according to preestablished licit operations) to output a new statement that in some appropriate way rearranges elements of the non-logical "content" contained in one or more of the premises.

In contrast to deduction, the conclusion of an abduction or abductive inference is not necessary. Instead, the conclusion of an abduction is explicitly an explanatory hypothesis that may or may not be true. There is, in Peirce's words, a "reason to suspect" that the hypothesis is true, but such suspicion is utterly different in spirit from the conclusion of a deductive inference. Whereas a deduction genuinely concludes, in the sense of reaching a definite terminus, an abduction remains intrinsically tendential and anticipatory. It is coordinated naturally with a further tendency to test and revise the conclusion in light of further experience, hence its broad alignment with the methods of experimental science.

In many respects abduction is closer in spirit to induction. Inductive inference is, generally speaking, less straightforward to describe than deduction, largely because the conclusion of an induction is not understood to follow necessarily from the premises. Typically, induction is more experientially oriented than deduction. A prototypical instance of induction is a series of particular observations from which some general claim is derived. For example, from the repeated sighting over several days of a point of regular brightness in the same part of the morning sky a short time before dawn, one (reasonably?) concludes that a similar event will take place tomorrow. Implicitly, the general claim "Such an event takes place every morning" is posited as the conclusion of the inductive argument.

Now consider the following example: from the repeated sighting of the bright point in the sky each dawn, one concludes that these points – daily encountered – are in fact visible effects of *one and the same object*, namely a neighboring planet in our solar system. What is the status of this line of reasoning? Rather than inferring a necessary or even determinately probable conclusion, this way of thinking introduces ("abduces") a new term into the local discourse of the implied "argument". We thus have two main threads of investigation:

1. Where does the new term come from? How is it introduced into the general logical grammar already structuring the premises?
2. What is the specifically *explanatory* relation of the new term to the terms already given?

Much of the scholarly interest in abduction stems from the breadth of its theoretical scope and the astonishing variety of its potential applications. Within the field of logic, narrowly defined, it presents a number of interesting formal problems. In philosophy of science, abduction plays the key role of capturing the internal conceptual process of hypothesis-formation, and it remains one of the least clarified aspects of scientific method. Concrete contexts in which theorists have applied models of abductive inference include law, medicine and engineering. Finally, research in abduction remains a central aspect of cognitive modeling, and holds particular relevance to constructive research in data analysis and artificial intelligence.

1.3.1 Early Characterizations

> Abduction consists in studying facts and devising a theory to explain them. Its only justification is that if we are ever to understand things at all, it must be in that way.
>
> Charles S. Peirce ([7], p. 205)

While there are certainly tentative approaches to the problem of hypothesis formation along the history of western philosophy, it seems fair to begin our survey of this problem with the work of Charles S. Peirce since it is with him in particular that the links of abductive inference with logical and mathematical formalism become first clearly thematized. Peirce himself stands at or very near the opening of modern formal approaches to logic, and his embedding of the problem of rigorous analytical notations for logic and mathematics in the more general framework of semiotics is one of the primary inspirations for the present work.

Of the many places where Peirce discusses abduction, one of the fullest general characterizations is found in his "Sundry Logical Conceptions" from 1903 ([7], p. 287):

> The whole operation of reasoning begins with Abduction, which is now to be described. Its occasion is a surprise. That is, some belief, active or passive, formulated or unformulated, has just been broken up. It may be in real experience or it may equally be in pure mathematics, which has its marvels, as nature has. The mind seeks to bring the facts, as modified by the new discovery, into order; that is, to form a general conception embracing them. In some cases, it does this by an act of generalization. In other cases, no new law is suggested, but only a peculiar state of facts that will 'explain' the surprising phenomenon; and a law already known is recognized as applicable to the suggested hypothesis, so that the phenomenon, under that assumption, would not be surprising, but quite likely, or even would be a necessary result. This synthesis suggesting a new conception or hypothesis, is the Abduction. It is recognized that the phenomena are like, i.e. constitute an Icon of, a replica of a general conception, or Symbol. This is not accepted as shown to be true, nor even probable in the technical sense, – i.e., not probable in such a sense that underwriters could safely make it the basis of business, however multitudinous the cases might be; – but it is shown to be likely, in the sense of being some sort of approach to the truth, in an indefinite sense.

In an abduction, one "concludes" with a kind of question and typically proceeds with a pattern of inquiry. As Peirce puts it, "The conclusion [of an abduction] is drawn in the interrogative mood" ([7], p. 287)

What then is abduction? Peirce's general schema for abductive inference runs as follows:

> The surprising fact C is observed. But if A were true, C would be a matter of course. Hence there is reason to suspect that A is true.

Expressed schematically, with a good deal of relevant information lost, Peirce's statement appears as follows:

$$\frac{\begin{array}{l} C \\ A \rightarrow C \end{array}}{A\ (?)}$$

Abduction looks, in short, like the deductive fallacy of "affirming the consequent". This resetting of Peirce's informal notion, versions of which are sometimes presented in the literature on abduction as if they are fully representative of Peirce's own, drops at least three important qualifications from the original statement, namely for Peirce:

1. C is not a sentence or proposition held to be true but an *observed fact*, that is, an experience involving dual aspects of, roughly, form (act of observation) and content (what is thereby observed).

2. The relation between A and C is both more and less rigid than the logical relation $A \rightarrow C$ of material implication. More, because it represents something like a counterfactual ("if A *were* true ..."). Less, because C is not said to follow necessarily or logically but merely "as a matter of course", certainly not as crisp a boundary as a typical logician might like.
3. The "conclusion" A takes the unusual form of neither certainty nor calculable probability but rather of *reasonable suspicion*, a subjective state the objective correlate of which is a licit conjecture.

Each of these three qualifications is in fact essential to abductive procedures. To elide any of them is to miss essence for accident. Nonetheless, if we take the simpler albeit less adequate formal scheme as a helpful starting point for grasping the method of abductive reasoning, it is clear that according to the scheme there are thus, roughly speaking, a pair of premises and a conclusion organized somewhat like a traditional syllogism.

Two elements of the schema stand out immediately. First of all, the characterization of the observed fact C as a surprising fact, which contrasts nicely with the claim that if A were true, C would be a matter of course. Even in the "premises" of the abductive argument what is at stake is a spectrum of expectation and novelty. Secondly, the type of conclusion is definitively non-definitive. What "results" from an abductive inference is less a claim then the practical orientation of "suspicion" which develops naturally into experimental activity and testing. In fact, as indicated above and as Peirce himself observed, the formal characterization of abduction is essentially the logical fallacy of "affirming the consequent". Yet a fallacy becomes an engine of discovery if it is treated not as something to be evaluated simply and immediately as either valid or invalid but rather as an internal deformation of the epistemic framework where it originated, a deformation that opens new possibilities and potentials.

Peirce's characterization calls to mind the formal schema later proposed by Carnap [8] that is meant to undergird both explanatory and predictive inferences:

1 $(x)\,(Px \supset Qx)$

2 Pa

3 Qa

Carnap [8] glosses his proposed schema as follows:

> First we have a universal law: for any object x, if it has the property P, then it also has the property Q. Second, we have a statement saying that object a has the property P. Third, we deduce by elementary logic that object a has the property Q. This schema underlies both explanation and prediction; only the knowledge situation is different. In explanation, the fact Qa is already known. We explain Qa by showing how it can be deduced from statements 1 and 2. In prediction, Qa is a fact *not yet known*. We have a law, and we have the fact Pa. We conclude that Qa must also be a fact, even though it has not yet been observed.

Carnap proposes here that the first premise, the expression of a general law connecting any particular case with a given property to a second property implied by the former, plays a distinctively explanatory role when the conclusion of the above syllogism is given as an experiential datum. In other words, from within a context in which one starts with the "conclusion" Qa given as a contingent fact, the premise Pa taken conjointly with the law-like premise $Px \supset Qx$ then serve to explain the fact Qa. Importantly, Carnap does not address the problem of how to generate the "law", that is, the (roughly speaking) conditional statement $Px \supset Qx$ as a tentative hypothesis in the first place. His point, rather, is simply that the two premises of the argument schema may be understood as an explanation of the conclusion in the sense that the epistemic status of the sheer fact Qa has by virtue of the theoretical introduction of the two premises been changed to that of a logical result. To translate this into the idiom of Peirce, the truth of Qa is no longer surprising since the conjoint truth of $Px \supset Qx$ and Pa converts Qa into a logically necessary conclusion.

Aside from unimportant technical details, Peirce and Carnap present essentially the same logical "picture" of explanation in terms of the parts out of which it is composed. They differ only in the order of the logical derivation itself, that is, the formal structure in and through which the component parts are coordinated into the unity of a logical argument. The different orderings given by Peirce and Carnap correspond to the difference between a logical fallacy and a valid deductive schema. Whereas the Peircean formulation follows the temporal order of surprising event, recognized (possible) explanatory connection, and hypothesis-as-tentative-conclusion and thus sacrifices the neatness of logical deduction, Carnap's sequence remains logically impeccable but temporally out of joint. The primary difference is that the Peircean "conclusion" is explicitly understood as a tentative hypothesis that as such must be posited, discovered or created as a real step in the process of inquiry such that it would, *if true*, explain the phenomenon. Carnap, on the other hand, refuses to view the explanatory function of $Px \supset Qx$ (taken together with the "occasion" Qa), as anything other than a simple given, very much in keeping with the logical empiricist dogma of a sharp distinction between the context of discovery and the context of justification.

The "covering law" model of explanation and, more generally, the logical empiricist program in philosophy of science is today no longer given much credence. In particular, the covering law model has failed to meet objections concerning the frequent two-way formulations of laws as biconditionals that nonetheless seem to resist two-way understandings of explanation (the argument from explanatory asymmetry) and the difficulty of distinguishing genuine laws of nature from conditionals that hold merely as matters of contingent fact (the argument from accidental generalizations). As an example of the former argument, consider a hypothetical law that coordinates, say, the size of individuals of some fixed species and their top speeds. Bigger animals of this species are, according to the conditional law in this case, faster. Let us assume further that this law is expressible as a biconditional. That is, the law states that any given size implies a determinate top speed and, conversely, any given top speed implies a determinate size. From a purely formal point of view, then, each of these sides of the law is equally explanatory of the other. Yet in this case in particular, while it seems plausible to conceive organism size as explanatory of speed, it seems unreasonable to think of an animal's top speed as explaining its size. What the example indicates is that explanation typically involves more than just a lawlike coordination. It also has a natural directionality that tracks real relationships of causality and explanatory power. The logic of explanation involves ordered hierarchies that may in some cases be independent of logical conditionals and biconditionals.

As an example of the latter argument from accidental generalizations, consider the "law" that the property of being a living being on the moon implies being a male human. So far as we know, this law is valid as a matter of sheer fact. But it clearly does not express anything like a rule of natural necessity. Standard expedients to avoid such merely accidental law-like conditionals, such as the use of counterfactual propositions, give rise to their own problems.

Such criticisms help us to see that the formal schema as presented above remains at the least less than sufficient for fully explaining the explanatory role itself of the inferred *explanans* in the schema. Rather than providing genuinely explanatory insight into what scientific explanation is and how it works, Carnap's formalism merely sidesteps the problem by redescribing it as logical subsumption under a conditional understood as a general law.

1.3.2 Recent Studies

Because of its central importance for practically all modes of scientific inquiry, abductive inference has been a well-researched and much-debated topic for both logic and philosophy of science in recent decades. Given this level of contemporary interest, it is itself a surprising fact demanding explanation in retrospect that abduction was *not* more of a subject of study prior to Peirce (despite certain indications in Aristotle) or even in the mainstream tradition of philosophy of science prior to the collapse of the logical empiricist program and the rise of Kuhnian paradigms, Feyerabendian anarchy and Stanford school pluralism in the second half of the twentieth century.

Much of the research into abductive reasoning has focused on the conditions of optimality that do or ought to constrain hypothesis selection from among possible candidates. Exemplary in this regard is the work of Peter Lipton in [9]. In such work, the *best* in the phrase "inference to the best explanation" bears the primary focus. To be sure, the application of naturally appropriate criteria such as simplicity of explanation, maximal conservation of prior theories, and so forth to given sets of relevantly probable candidate hypotheses is one of the major facets of abductive inference.

An overview of the breadth of the terrain named by the deceptively simple term "abduction" and a critical evaluation of many of the theoretical attempts made to map that terrain is available for the interested reader in [3]. There, Gabbay and Woods provide a formal scheme that is meant to capture the key characteristics of abductive reasoning as expressed by a set of five premises entailing a pair of conclusions. The component building-blocks of the premises and conclusions are the following variables:

- T: a cognitive *target*, the desideratum of the abductive process
- K: a (present) knowledge-base
- K^*: any future knowledge-base presumptively accessible from K
- R an "attainment relation" specifying which knowledge-bases K hit some target T
- R^{pres}: R as *presumed* to hold by the relevant epistemic agent
- H: hypothesis
- $C(H)$: H as held to be *justifiably conjecturable*
- H^C: the *discharge* of $C(H)$ as actually and appropriately conjectured

With these rough and admittedly presumption-heavy materials in view, what they call the GW-schema of abduction may then be expressed as follows ([3], p. 47):

1. $T!$ [declaration of T]
2. $\neg\ (R(K,T))$ [fact]
3. $\neg\ (R(K^*,T))$ [fact]
4. $R^{pres}\ (K(H),T)$ [fact]
5. H meets further conditions $S_1, \ldots, S_n$ [fact]
6. Therefore, $C(H)$ [conclusion]
7. Therefore, H^C [conclusion]

As in any formalization, what matters are the relations among the parts. For this schema, the key problem of abduction arguably resides in step 4. The presumptive attainment relation that connects the knowledge-base-as-revised-by-H to the epistemic target T is understood here as a sui generis type of relation that characterizes abductive reasoning as such.

In this respect Gabbay and Woods are critical of what they see as an overly narrow conception of abduction characterized by (in their terms) *explanationism* and *consequentialism*, particularly the latter. For this reason, they reject (albeit with much appreciation) the formal scheme that they propose as representing the current standard model that broadly aligns such theorists as Aliseda, Kowalski and Kuipers, and Magnani (hence their designation of it as the "AKM-schema"). See [3], p. 48-9, formalism slightly modified:

1. E
2. $\neg(K \implies E)$
3. $\neg(H \implies E)$
4. $K(H)$ is consistent
5. $K(H)$ is minimal
6. $K(H) \implies E$
7. Therefore, H.

Overall, [3] provides an invaluable compendium of the state of the art in formal approaches to abduction. Especially noteworthy is its breadth of coverage, which offers detailed reconstructions of even the methods and positions that are finally criticized or rejected by the authors. It is in this respect a useful text for charting the predominant "space of reasoning" within which theories of abductive inference are currently being proposed and reviewed from various quarters. Among other things, the authors rightly point to the centrality of what could be called the "framing problem" for theories of abductive inference. As they put it, "how much of abduction to try to capture schematically is an important question" (p. 53), and, we might add, a largely open one in which for example the further claim that even "omissions of abductively salient factors need to be justified" (ibid.) still presumes that the theoretical coordinates defining such "abductively salient factors" are already intuitively available.

Part of what makes abduction especially problematic for any attempt at rigorous investigation is the intrinsically hazy nature of the object-matter itself (the Husserlian *Sache selbst*) that is in fact meant to be investigated. It seems that the very posing of the problem of abduction requires something much like an act of abduction as a veritable condition of possibility. We would like in the present work to focus on this as it were second-order problem of abduction even more doggedly, even going so far as to suggest that this "self-inclusive" logic might be of abductive reasoning's very essence.

Working in the more restricted space of formal propositional and first-order logic but still aiming to capture the essential dynamics of abductive inference is the work of Aliseda [10], to which we will return in chapter three. Making extensive use of the technique of semantic tableaux, Aliseda examines in particular how criteria used to shape the space of admissible hypotheses correspond to variable strategies for enriching and restricting the class of satisfiable models under consideration. Aliseda's methods have been influential for the present approach particularly due to the semantic or model-theoretic perspective on formal languages that they use to parameterize abduction. Even if our approach substitutes a more general and broadly naturalized point of view on the syntactical side of abductive inference via Peirce's notion of iconic semiotics, Aliseda's standpoint on abduction has strongly influenced our own.

Additionally, Magnani's extensive engagement with the problem of abduction spans [4, 5, 11, 12]. In lieu of any strictly formal approach that would tend to restrict the "matter" of abductive inference to truth-functional representations, Magnani enlarges the scope of abduction to encompass a wide of variety of cognitive environments. His name for this approach is "eco-cognitive" reasoning. The strength of treating abduction in this broader frame with its more flexible range of cognitive activity is, from the present point of view, that the highly particular use of abductive reasoning in scientific inquiry becomes more clearly indicated as a special case of the more extensive phenomenon. The interesting questions that thereby arise concern the specific conditions that must be met to further qualify abductive reasoning as appropriately scientific.

Finally, from a computer science perspective recent advances in the field of *Logic Programming* seem to support the notion that abductive inference, regarded as a procedural semantics, finds natural models in the structure of programming languages based on non-classical logic. For a comprehensive introduction to this aspect of the employment of logic programming as a representational and reasoning tool see [13].

1.4 Abduction and Context-Bound Reasoning

Perhaps the most intriguing feature of abduction is its natural encampment within real-world processes of deliberative and ethical reasoning (see the approach of Harman in [14]). Anyone who has ever sat through a committee meeting charged with the task of coming to some controversial decision knows that the necessary conclusions reached by formal deductive logic are typically not relevant to the actual interplay of proposals and evaluations made by the parties concerned. Instead, discussions tend to take place in the uncertain and interrogative subjunctive: what if ...? While strictly speaking, the consequences of such a "what if ..." are subject to deductive structures of implication (perhaps understood loosely), the posing of the counterfactual "what if ..." itself may be understood rather as a type of abduction (see [15]). Similarly, any positive proposal made under conditions of relative uncertainty stands as a kind of hypothesis, and well-justified proposals are of course typically coordinated with accurately explanatory (successful) hypotheses ("If we follow X plan of action, goal Y will be achieved as a matter of course").

In fact, abduction is always a key element of reasoning involved in the need for decision and action under conditions of relative ignorance. In this respect, the importance of understanding and optimizing abductive inferences may be signaled by mentioning the current urgency of responsible policy-making with respect to what is quite possibly the most far-reaching and crucial abduction ever made in human history: imminent global climate change.

What we are especially at pains to emphasize in this regard is the real, context-bound character of abduction. Abductions not only *are not* but, in our view, *cannot be* made arbitrarily. At best, an arbitrary inference of this type is a quasi- or pseudo-abduction. Any genuine abduction takes place in a concrete context of knowledge, ignorance and experience. Each of these three elements is not merely a necessary condition for abductive inference, but more strongly each of them enters into the process of abductive inference as an essentially dynamic component. To begin with *knowledge*, abduction is strictly meaningless outside of some context of relatively established epistemic "bearings" with respect to which the surprising event may be registered as, precisely, surprising. There is no reason to restrict this notion of knowledge to some particular theory of scientific knowledge-bases or *Weltanschauungen*. Our commitment to the claim that some antecedent epistemic context is an essential component of abduction does not imply a commitment to any particular concept or theory of what knowledge is. On the contrary, it seems that the most robust characterization of abduction will require a conception that ranges over a wide variety of knowledge-types.

Knowledge alone is not sufficient, however. It is necessary also to introduce *ignorance* as an equally essential component of abduction. It might seem that this notion would come, as it were, for free with any given characterization of knowledge. Ignorance, after all, is simply the contrary of knowledge. The ignorance of some epistemic agent consists in whatever is not known. However, it is clear upon reflection that in the actual context of inquiry of whatever sort, ignorance is not simply a privation of knowledge, but is – or at least can be – a goad to discovering or generating knowledge. On the difference between one's merely not-knowing-*X* and one's knowing-that-one-does-not-know-*X* depends the possibility of actively seeking and not merely stumbling upon truth. This point may appear to be a minor quibble, but in fact it has profound ramifications for theorizing abduction.

Finally, abduction becomes intelligible only in a concrete context of *experience*. Just as with knowledge, we are not committed a priori to any particular concept or model of experience, but we insist that abductive reasoning is necessarily experientially grounded and processed. The basic "shape" of experience as a cognitive agent's immediately affective and mediatedly epistemic relation to the real world is the warp to abductive reasoning's woof.

In this respect, our approach to abduction is in broad accord with Peirce's philosophical method of pragmatism. In sharp contrast to Descartes and the dominant modern subjectivist tradition in philosophy, Peirce rejects the method of radical doubt that would aim at establishing philosophical foundations in the apodictic nature of consciousness through the method of radical doubt. Not only does Peirce deny the utility and philosophical-scientific fruitfulness of proceeding via radical doubt; more pointedly, he denies that such doubt is even possible.

From this point of view, the validity of abduction thus depends on semantic as well as syntactical considerations. It is in this respect closely allied with a variety of philosophical problems concerning linguistic meaning and interpretation. In particular, we point to the work of philosopher Donald Davidson as a helpful illustration. Davidson has developed a fruitful theory of meaning that builds upon Tarski's conception of truth. For Davidson, meaning is parasitic upon truth, and therefore the common-sense or "folk" theory of language that would understand a given sentence *S* primarily as possessing a particular meaning *M* that would in some given secondary context be judged properly to be true or false depending on whether the state of affairs expressed by the meaning in fact obtains in the relevant context is itself an artifact of certain misconceptions about how language actually works. In Davidson's view the most important questions about language's power to signify reality (and thus to be either true or false) are investigated through inquiry into the possibilities for and obstacles to *translation* from one language to another. This emphasis on translation *across* languages does not annul but rather helps to explain the fundamental relationship between language and world. In [16] he applies this far-ranging philosophy of language to the problem of generating interpretations of linguistic *mis*-usage and singular, non-standard expressions in an especially illuminating manner.

Davidson initially characterizes the notion of "first meaning" in language, that is, roughly, the "normal" meaning expressed by a linguistic utterance. The three basic characteristics of "first meaning" are as follows:

1. It is systematic.
2. It is shared.
3. It is governed by learned conventions or regularities.

He then asks how these three characteristics or conditions are to be modified when it is a question of *malapropisms*, or non-standard (but still meaningful) *mis*-uses of language. He introduces the distinction between a "prior" and a "passing" theory of meaning, the former being the background understanding of "normal" language use and the latter being the ad hoc construction of a theory of what *might* be meant by a given speaker under concrete contextual conditions. Clearly, it is the formation of a "passing" theory in Davidson's sense that allows language-speakers to make sense of malapropisms and, in general, non-standard usages of language that are nonetheless comprehensible.

Davidson's overall conclusion is that although it becomes possible on this basis to make sense of two individuals "having the same language", it is impossible to identify precisely just what it is that they have. All that can be specified is that whatever it is that the two speakers have, it is relevantly "the same". We will see how this apparently strange if not contradictory notion of "sameness" is captured formally by the categorical notion of isomorphism. Yet even if the essence of language itself cannot be determined, according to Davidson the *use* of language may still be quite nicely characterized (p. 106):

> But we might try to say in what a person's ability to interpret or speak to another person consists: it is the ability that permits him to construct a correct, that is, convergent, passing theory for speech transactions with that person.

Finally, Davidson points out that this more pragmatically grounded conception of language ultimately puts language in continuity with other forms of practical engagement with the world. There is thus no longer a clear division between concrete interaction with the world and theoretical (linguistic) representation of it. Davidson summarizes (p. 107):

> [W]e have abandoned not only the ordinary notion of a language, but we have erased the boundary between knowing a language and knowing our way around in the world generally. For there are no rules for arriving at passing theories, no rules in any strict sense, as opposed to rough maxims and methodological generalities. A passing theory really is like a theory at least in this, that it is derived by wit, luck, and wisdom from a private vocabulary and grammar, knowledge of the ways people get their point across, and rules of thumb for figuring out what deviations from the dictionary are most likely. There is no more chance of regularising or teaching the process of creating new theories to cope with new data in any field - for that is what this process involves.

Abduction works in the same way. It is always local and there is no need for any particular abductive inference to affect the global structure of knowledge. Nonetheless, the global epistemic framework does over time shift in accordance with the unfolding of originally germ-like abductive events. It would appear to be greatly advantageous for the development of both science and the philosophy of science if the dynamics of this alternately punctual and glacial process could be formalized.

1.5 The Problem of Formalization

The question of whether or not abduction can be formalized has been historically controversial [17]. Why does formalizing abduction raise specific theoretical problems and concerns? Most importantly, there is the problem of "creative novelty" in abductive inference. Unlike both deduction and induction, in the conclusion of an abduction a non-logical term typically appears that appears nowhere in the premises. This would seem to obviate the sort of rule-bound, algorithmic framework that makes automated procedures for deductive inference possible. In particular, paradigmatic events of abduction usually rely on the non-formal and profoundly opaque "databases" that actual human beings tend to inhabit. When Archimedes shouted "Eureka!" it was not because he had finally concluded a tedious calculation, but rather because the solution itself "surprised" him, percolating up perhaps from dark and ambiguous experiences of swimming, childhood toyboat play and so on.

Related to the problem of creative novelty is the impossibility of a definite form/content distinction within processes of abduction. Abduction is not an abstract activity, but a definitively situated one. *Someone* abduces, from within the limits and real constraints of an epistemological (and not only epistemological) situation. In short, what is at stake in abduction is not simply epistemology, but epistemic ontology. The relevant aspects of abduction include not only the abstract states of factual and theoretical knowledge, but the very being of the epistemic agents involved: who are they? what kind of animal (or machine, or alien intelligence)? how do they live? how do they think? why do they care to know?

In this respect, theorizing abduction requires that attention be paid to the real dynamics of the strategic dimension of epistemic inquiry. Abduction is intrinsically economic in the sense that it becomes intelligible only within the framework of real chance, risk and possible reward that is the very mark of actual existence (and actual science).

In short, formalization (implicitly or explicitly) looks for the essence of a phenomenon by picking out what is invariant within it. This understanding of form as essence is at the heart of the philosophical tradition of knowledge stretching back at least to Socrates and Plato. Yet abduction is intrinsically situated, contextual and variable. It exists only in respect of actual, existentially situated problems that might be conceived and solved in multiple ways. Abduction thus invents itself as it is repeatedly instantiated – it is simultaneously a what and a how question: What explains X? and How do I find the "what"? Indeed the distinction between what is looked for and how it is sought provides a helpful way to classify the three types of inference: in deduction the "what" is an implicit consequence and the "how" is encapsulated in formal rules; in induction the "what" is some general law and the "how" is a selection of relevant features and an extension of purview to include all (or most) of some type; in abduction, the "what" is an explanatory hypothesis and the "how" is the very problem itself.

Rather than being subject to formalization, then, it seems that an account of abduction is necessary to explain the formal itself. The formal is always embedded and practiced in some relatively concrete (that is, less formal) ambient context. It requires the real as what is always already subject to possible formalization, and yet necessarily precedes formalization. This is why otherwise interesting attempts to develop a formal logic of abductive reasoning, as in [18], remain strictly exterior to the real phenomenon of abductive reasoning.

Abduction always involves the difference measured between an actual state of surprise and a potential state of (possible) explanation and understanding. In general, the change of state is here of greater interest than either of its endpoints. Learning and forgetting are two typical forms of transition from one state of knowledge to another. Such transitions are the inevitable lot of epistemic agents situated in a world that changes and with the constituent objects and processes of which those agents must interact in both familiar and novel ways. Even if truth is eternal (a dubious proposition, we know), states of knowledge most certainly are not. Knowledge is localized, temporalized and grounded in controlled interaction with the ambient environment (which extends indefinitely). We presume that epistemic agents are also economic agents, in the sense that they must make choices among competing possibilities without being guaranteed of the consequences of their choices.

Logical inference, too, may be characterized as a type of transition from one state of knowledge to another, although in the case of inference it is not a matter of localized interaction with the world provoking the transition, but of interactions among already constituted ideas. Whatever the status of such ideas – a notoriously thorny philosophical question – the outcome of inference is more of the same, although amplified. This is to understand logical inference as an event. Similarly, observations and measurements are always epistemic events. The very concept of measurement presupposes a context of knowledge and a passage from relative ignorance to relative knowing.

The ground of abduction is the surprise registered by an agent in the face of some phenomenal datum. It is important to distinguish this component of surprise from the content of the phenomenon itself. It is not the event's intrinsic features that make it surprising, but the relation – or rather, the lack of relation, the failure of correspondence – between the event and the theoretical framework of explanations and expectations through which the agent comports herself in the world.

One final challenge for formalizing abduction is thus how to register this element of "surprise", obviously not only as a psychological experience but rather as an objective feature of data relative to some theoretical context. Clearly, it is possible to define "surprise" negatively relative to some given knowledge database or subjective perspective. But a thorough investigation of abduction also requires us to ask: Are there *objectively* surprising phenomena in nature itself? Answering this question in the affirmative is tantamount to affirming objective "laws" or "structures" in nature that are not, however, deterministic in a universal and univocal way. One strong candidate for such an objectively surprising phenomenon is the accord of physical and mathematical laws. In general, good abductive inferences in science must follow the grain of such objective patterns of regularity and exceptionality. Surprise from this perspective becomes then a necessary but not sufficient condition for scientific abductive reasoning. Necessary too is the selection of relevant or salient surprises, that is, surprises that themselves promise explanatory insight into nature. Many surprising phenomena, after all, do not provoke abductions but merely mark intermittent discontinuities in attention and theoretical shrugs. Often – more often perhaps than not – things "just happen" and are registered as no more than brute facts.

Indeed, one of the greatest challenges for formalizing abduction is to bring both the factual and normative dynamics of this selective process into the formal model.

One final note: all of the preceding may appear to have been couched in a deliberately psychologistic way. This is not because we intend to restrict abduction to the human or conscious realms. Quite the contrary. Our aim is to develop a theoretical framework for abduction that is general enough to include artificial intelligence, robotics, non-human animals, etc. Ordinary human cognition will be a privileged source of instances of and insights into abductive processes, but this is by no means meant to exclude non-human and non-conscious agents.

1.6 The Generic Shape of Abductive Reasoning

Abduction depends on a fundamental duality of knowledge and truth: knowledge tends to extend outward and is essentially conservative; truth tends to develop through more and more finely grained determinations. Knowledge accumulates, truth specifies. Framed in this way, the basic shape of abduction appears as that of a three stage process:

(1) *A surprising event.* The first stage of the abductive schema involves a phenomenal given. Something occurs or appears in a particular existential/theoretical context that is sufficiently unexpected that it poses a "problem" of explanation.

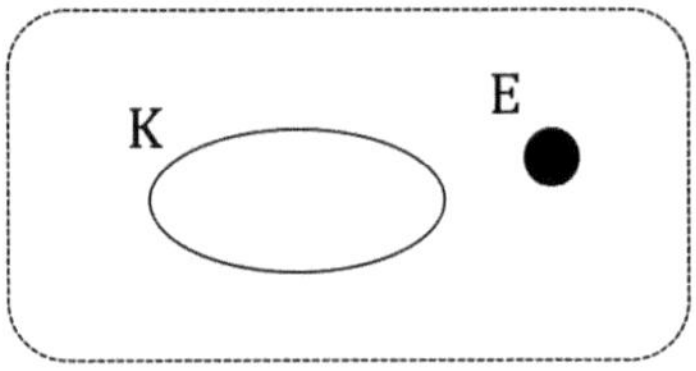

An informal yet suggestive way to represent this stage is to picture the existential context as a space M of factual givens (the ground model), the theoretical context as a projection into this space of a distinction between what is explained and what is not thus delimiting a subspace $C \subseteq M$ (what of M is theoretically constructible) of what is explainable by the current theory, and a "part" E in the space outside of C that represents the surprising event as a subset of the complement of C in M. As a helpful toy model intuition of this representation, we picture the space M as the plane with a specified knowledge-base K playing the role of the subspace C as a closed curve in the plane, and the event E as a point or germ-like spot in the plane outside of K.

(2) *The expansion of cognitive possibilities.* The second stage of the abductive schema involves a controlled relaxation of the limits of the theoretical context K, allowing the space of possible theoretical explanations to vary. This may occur either ampliatively by extending the reach of K so that it contains E or subtractively by removing those elements of K that make E unexpected or surprising. The class of these possible extensions we designate H_n. Important point: typically, this expansion

is neither fully determinate nor absolutely undetermined. It generally constitutes a structured "space" of possibilities.

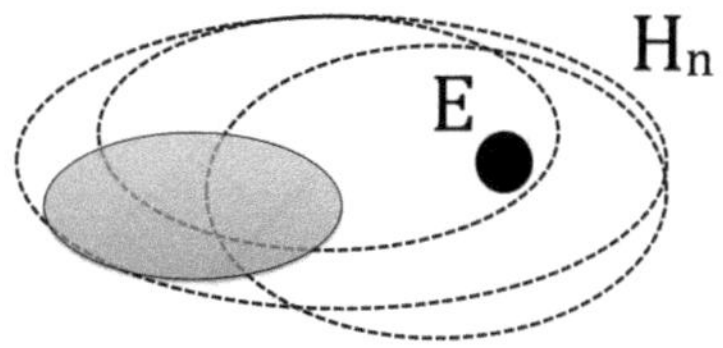

(3) *The contraction/collapse of the enlarged space of possibilities*. Finally, the third stage of the schema regulates the "collapse" of the expanded space of explanation to some intermediate position between the original context and the enlarged space – specified by either a single, determinate difference between the original and the new theoretical contexts (a definite hypothesis) or a range of differences (a class of potential hypotheses or a variable type of hypothesis). We designate this new conjectural state $K[H^*]$.

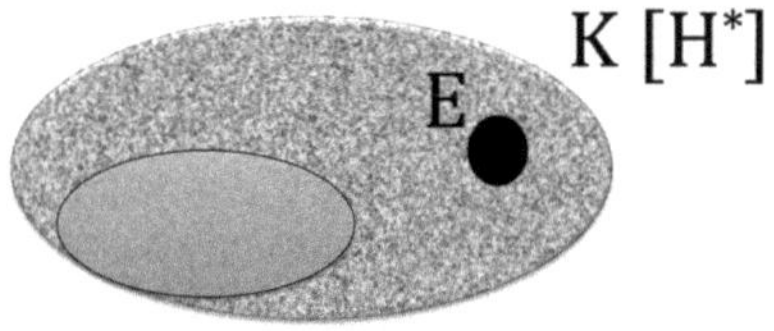

Consider the following informal illustration of this three-stage process of initial surprise followed by the expansion and then the contraction/collapse of epistemic possibilities. Note how this pictorial model potentially coheres with the figures presented at the beginning of the present chapter used to illustrate the contravariant relationship between sequences of knowledge-states and potential models of reality consistent with those states.

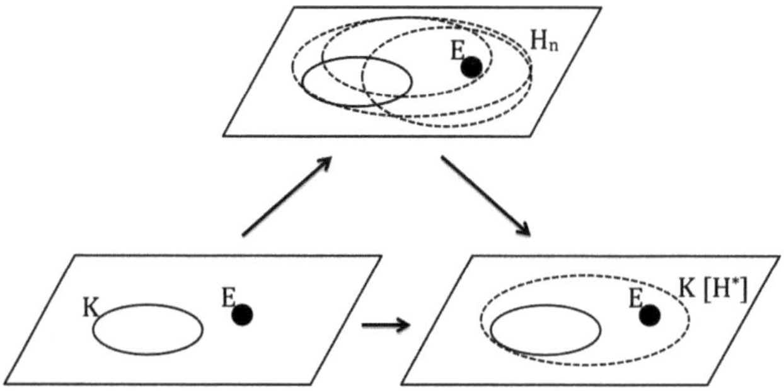

The essence of the categorical approach that we will be following throughout the present book is to abstract relatively from the aggregate data distinguishing particular objects at whatever level of theoretical consideration and to attend rather to the systems of relations that coordinate those objects. In the case of the diagram of abductive "shape" just proposed, it may be helpful to forget the internal content of each of the three stages involved and to shift attention to the transformations from one stage to the next as indicated by the arrows. We thus allow the three inscribed planes to collapse or "contract" to indistinguishable dots and provide labels to determine the arrows instead.

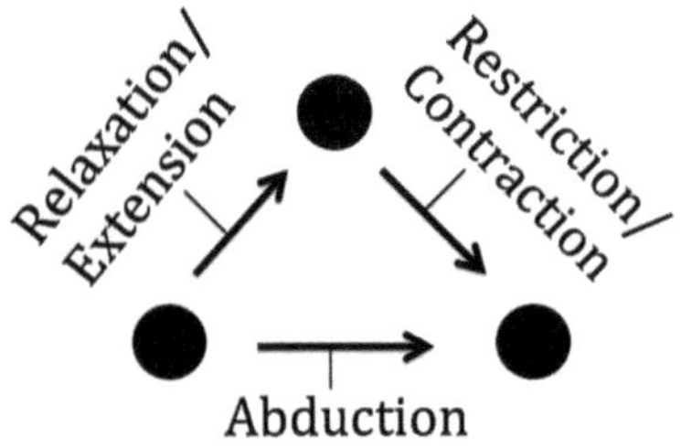

This diagram is meant to suggest that the relevant "shape" of abductive reasoning is in fact a composition of two sequential and dually oriented relations, the first of the relaxation of formal strictures and the concomitant extension of a space of possibilities and the second of the controlled restriction of that enlarged space and its contraction to some selected subregion. Abduction "is" from this point of view the relation between two states that results from a theoretical extension of possibility due to the relaxation of constraints followed by a critical restitution of constraining form that adjusts to some appropriate criterion or criteria.

Perhaps the simplest cases of abduction are found when the distinction between the existential and theoretical contexts is not rigidly drawn. In such cases, the surprising event E to be explained is treated as more or less independent of any particular epistemological context. The difficulties involved in deciding how best to adjust an already existing network of theoretical commitments to accommodate the new datum may be effectively set aside. The event E is simply given as such as an unexplained fact, and the abductive task is to posit a local explanatory context that would, if indeed true, reduce the surprising character of E to mere quotidian predictability.

Example 1 Peirce outlines the logical structure of what he calls "perceptual abduction". This is the experience, so ordinary as to be practically invisible, of perceiving something as something, for instance seeing a collection of black marks on a white page as (simultaneously recognized yet nonetheless compositionally ordered) letters, words, sentences, a text. Peirce (cited in [19], p. 91) summarizes:

> The form of the perceptual abduction is:
>
> A well-recognized kind of object, M, has for its ordinary predicates $P1$, $P2$, $P3$, etc., indistinctly recognized.
>
> The suggesting object, S, has these same predicates, $P1$, $P2$, $P3$, etc. Hence, S is of the kind M.
>
> The first premiss is not actually thought, though it is in the mind habitually. This, of itself, would not make the inference unconscious. But it is so because it is not recognized as an inference; the conclusion is accepted without our knowing how.

It is then possible to distinguish two poles or types of abductive inference. One pole is that of relatively decontextualized theory-formation. Given a set of data essentially independent of some explicit pre-given theoretical context, what hypothesis or (small) collection of hypotheses would, if true, explain the data? This type of abduction puts greater weight on the explanatory element of the inference. In effect, the burden of the inference is put on the second premise in Peirce's earlier schema. The first premise, at the limit, reduces to a simple claim: X occurs, or X is given. The surprise is reduced to mere factuality.

The other pole of abduction is approached as greater and greater emphasis is placed on the surprising character of the event E in the first premise. For abductions at this pole, the main challenge is less to explain E than to accommodate E to an already established epistemological framework. Here, the abductive problematic is defined most essentially in terms of a tension between an expansionist, inventive tendency that aims to proliferate possible explanations, to maximize the inclusion of as many possibilities as possible, so as to be most likely to hit on the truth, and a conservative tendency that aims to preserve as much of the already established epistemic context as possible and to alter it, if necessary to account for the phenomenon at stake, as little as possible.

1.7 Mathematical Abduction in Science

The present approach is especially concerned with the role of abductive inference in scientific reasoning. The above discussion should make it clear that abduction is a very broad category that includes reasoning in ordinary experience and a variety of non-scientific contexts. Although focusing on scientific reasoning draws a rather narrow circle with this immense field, the relations between this circumscribed arena and the much larger territory that surrounds it should remain in view, peripherally, as a constant horizon of intelligibility. Approaching scientific reasoning from this perspective emphasizes the continuities between scientific and other forms of cognition at the same time that it highlights the special characteristics that distinguish natural scientific inquiry from other domains of abductive reasoning such as ordinary perception and motility. Among such characteristics, the paramount importance of mathematics throughout natural science stands out as possessing profound epistemological significance.

The *locus classicus* of the early modern revolution in science is often held to be Galileo's 1623 *Il Saggiatore* (*The Assayer*) and its famous statement that Nature is a Book written in the language of geometrical mathematics ([20], p. 183):

> Philosophy is written in this all-encompassing book that is constantly open before our eyes, that is the universe; but it cannot be understood unless one first learns to understand the language and knows the characters in which it is written. It is written in mathematical language, and its characters are triangles, circles, and other geometrical figures; without these it is humanly impossible to understand a word of it, and one wanders around pointlessly in a dark labyrinth.

The most useful theoretical tool we possess for translating the experiential and observational data of scientific experiment into a register of comparison and exactitude is of course mathematics. Measurement is the general method employed by experimental science to translate qualitative observational data into quantitative mathematical data. The very possibility of measurement typically depends upon a variety of background assumptions that are typically unthematized as such. Only under exceptional circumstances do the background conditions that make measurement possible rise to the level of theoretical reflection. Usually a scientist is able to take for granted, for example, that rulers are rigid and their lengths remain invariant under motions through space. However, it was precisely by calling this presupposition into question, that is, bringing it to the foreground as a presupposition so that it might be considered a variable parameter in the epistemic field, that Einstein was able to formulate the Special and General Theories of Relativity. In many important cases of abduction, the core of the "explanatory hypothesis" is not so much the positing of a positive law but rather something more like the relaxing of an axiom, allowing a fixed presupposition or epistemic habit to vary.

What seems to make mathematics uniquely important in this regard is its independence in principle from any natural or empirical determinations. This conclusion has something of the appearance of a paradox. Mathematics is such a useful tool for modeling the natural world with exactitude (as well as offering a generic space for comparing and contrasting results from its heterogenous domains and seeking possible explanatory hypotheses for it) precisely because mathematics is independent of that world.

Bookstein in [19] examines in detail the de facto role and de jure appropriateness of "numerical inference" in scientific inquiry and particularly in the forming and testing of scientific hypotheses. A crucial element of Bookstein's approach to mathematical abduction in science is his coordination of abductive inference with the notion of *consilience*, that is, the tendency of scientific inquiries originating in diverse domains or oriented according to different perspectives to converge on truth in a unified and unifying way, as for instance in the synthetic "meeting up" of the principles and methods of chemistry and biology in molecular biology and genetics.

Bookstein makes the case that what he calls "numerical reasoning" is the best guarantor of consilience among the diverse sciences. The term consilience entered the lexicon of philosophy of science originally with William Whewell, an important influence on Peirce, and served as the title of and main theme of Wilson's 1998 book [21]. Bookstein's own focus is on consilience via quantitative analysis ([19], p. 26):

> Consilience in the natural sciences is the assumption that measures of the same quantity by otherwise different machines or experiments should result in the same numerical patterns.

Bookstein points out that using numbers is scientifically effective both for "condensing knowledge", that is, representing complex states of affairs in compact, easily manipulable ways, and also "*acquiring* such knowledge", namely, enabling more or less exact measurements of various kinds (p. 31). The relationship itself between these two scientific roles of number – representation and measurement – is itself a crucial aspect of science. The importance of this relationship is perhaps more clearly seen in the case of scientific instruments, which must both precisely and accurately measure relevant data *and* make those measurements communicatively available for the appropriate audiences (consider the now typically taken for granted digitization of all forms of scientific data and the rise of computational methods even inside "pure" mathematics). Bookstein writes (ibid.):

> Hence part of the authority of numbers arises from the reliability of the machines that generate them, and the foundation of any numerical science rests no more on the "testable laws and principles" than on the machines by which we produce the numbers in terms of which those laws and principles are phrased.

As treated also in Galison [22] in the context of subatomic physics, the interplay in this regard of experimental, theoretical and rhetorical/argumentative methods in science provides an essential background framework – a kind of transcendental *a priori* infrastructure – for doing science and understanding its results. We argue that it is the intrinsic iconicity (a term to be defined more precisely in chapter two) of mathematics that explains why it and the forms of "numerical reasoning" that follow from it are so deeply embedded in this core scientific infrastructure.

1.8 Conclusion

What is abduction? Abduction is a transition from an epistemic state of reflective ignorance to one of contingent hypothesis. To be in a state of reflective ignorance is not simply to lack knowledge but, in some non-trivial sense, to know that one lacks knowledge. In the case of abduction, reflective ignorance is here understood to mean reflective ignorance about some surprising event, something that takes place or happens. Abduction is a response to something E, some event, occurring in some family of contexts C that generates a hypothesis H that, if it were true, would explain E. To say that H would explain E if H were true, is to say that H is such that E loses its surprising character in light of H's real possibility, in Peirce's terms were H true, E would then "follow as a matter of course."

An abduction is not like solving a problem. It is rather like locating a direction in which the solution to a problem seems likely to be found. It is hopeful, but not too much. It reserves energy, as it were, for testing and trying. When the dynamics registered by the basic shape of abduction are situated in the general epistemological context with which the chapter began (the duality of knowledge's extension and reality's restriction), it becomes clear that a key element of abductive inference, and part of the reason why such inferences resist both formalization and automation, is that abductions must typically take into account not only a single "one-step" transition in a theoretical context, but a comparison among indefinitely many, indefinitely long "paths" through a space of potential epistemic states.

There are then two core issues for abduction and formalization as presented above. The first is that the formal schema fails to account for the actual explanatory power of the conditional (or quasi-conditional) relationship between the surprising event and the hypothesis. It would seem that any such explanation requires more detailed information about both the event and the hypothesis. It is necessary to "look inside" the concrete objects denoted by the formal indices of E and H. The real and not merely formal implication (or quasi-implication) of the one by the other depends upon concrete features of each of them that are simply unavailable outside of the concrete, experiential encounter with the event that in fact spurs inquiry, an encounter that consists of both real objects and causal relations in the experiential environment and actual features of the epistemological state and background of the inquirer. The "grain" of logical analysis – however fine – is always coarser than the real itself. Without any means for exhibiting and examining the internal structure of the semantic objects really denoted by the variables of the formal syntax, there is no way to account fully for the internal dynamics of the inference. Abduction is a semantically grounded and not merely syntactically structured form of reasoning. This is just another way of expressing the fact that abduction takes place in the real world, in a context of real decisions, real risks and actual limitations, not formal representations of these. While there may very well be something like a general grammar of abductive inference – certainly the formulations by Peirce, Gabbay and Woods, Aliseda and others suggest as much – any such grammar will remain insufficient for detailing the actual mechanism of successful abduction in any particular case, unless it maintains some way of building in the relevant information in an appropriately layered and compositionally structured fashion.

The second problem runs deeper. The actual process of abduction requires, at least in many cases, a genuinely inventive leap that cannot be taken into consideration prior to its actual articulation. In fact, in many cases the "success" of an abductive inference consists precisely in the adjustment of an already extant theoretical framework that must be altered in a way that prior to the abduction itself would be deemed impossible. This may take place either through amplification or reduction. The scientific revolution itself required, for instance, the subtraction at least to some extent of a providential divine will from the theoretical context of the Middle Ages. Abduction thus takes place both within a given epistemological context and simultaneously in an expanded context that includes possible variations on the original context that cannot be foreseen in advance of the abductive inference itself. Any formal modeling of abduction that simply builds in the expanded context in advance will inevitably leave unaddressed the problem of how the context itself expands and contracts within the actual process of abduction. Abduction is both inside and outside the limits of an actual epistemic situation. It thus seems that any formal characterization of abductive inference faces some variant of the paradoxes that ensue when a formal system is required to refer to its own totality and its own limits. This quite general problem is what Graham Priest has analyzed in terms of what he calls the "inclosure schema" [23].

The correct conclusion, however, is not a negative one. It is not a matter of aiming to establish a quasi-Kantian critical limit concerning abduction and its formalization beyond which speculation would be vain. It is a matter, rather, of suggesting that the type of formalization endemic to classical logic in particular may be inappropriate for abduction. We propose that the mathematical structure of presheaves and sheaves provides a versatile framework for addressing both of these issues. In particular, the structure of presheaves provides a natural mathematical setting for the kind of model-based or diagrammatic reasoning that characterizes experimental science and that seems in actual practice to provide the best avenue for treating the concrete features of the surprising event to be explained as positive data with intrinsic relationships to possible explanatory hypotheses. We examine this aspect of abductive reasoning in the following chapter in terms of the Peircean semiotic category of iconicity. In addition, working with the mathematical context of presheaves and sheaves suggests at least some insights into the problem of the creative character of abduction. We look into this aspect especially in chapters four and five, which examine Badiou's use of set theory as a mathematical model-space for philosophical ontology and which take the generic in forcing and the sheaf construction over a complete Heyting algebra as regulations of real determination, hypothetical freedom and epistemic coherence.

References

1. D. Walton, *Abductive Reasoning* (University of Alabama, Tuscaloosa, 2005)
2. T. Bertolotti, *Patterns of Rationality: Recurring Inferences in Science, Social Cognition and Religious Thinking* (Springer, New York, 2015)
3. D.M. Gabbay, J. Woods, *The Reach of Abduction: Insight and Trial* (Elsevier, Amsterdam, 2005)
4. L. Magnani, *Abductive Cognition: The Epistemological and Eco-Cognitive Dimensions of Hypothetical Reasoning* (Springer, New York, 2010)
5. L. Magnani, *Abduction, Reason and Science - Processes of Discovery and Explanation* (Springer, New York, 2000)
6. P.A. Flach, A.M. Hadjiantonis (eds.) *Abduction and Induction: Essays on their Relation and Integration* (Kluwer, Dordrecht, 2000)
7. C.S. Peirce. Ed. Peirce Edition Project (1998). The Essential Peirce: Selected Philosophical Writings, Volume 2 (1893–1913). Indiana University Press: Bloomington and Indianapolis
8. R. Carnap, in *Philosophical Foundations of Physics*, ed. by M. Gardner (Basic Books, New York, 1966)
9. P. Lipton, Inference to the best explanation, *A Companion to the Philosophy of Science* (Blackwell, Malden, 2000), pp. 184–193
10. A. Aliseda, *Abductive Reasoning: Logical Investigations into Discovery and Explanation* (Springer, New York, 2006)
11. L. Magnani, The eco-cognitive model of abduction. J. Appl. Log. **13**, 285–315 (2015)
12. L. Magnani, The eco-cognitive model of abduction II. J. Appl. Log. **15**, 94–129 (2016)
13. L.M. Pereira, *Reasoning with Logic Programming*, vol. 1111, LNAI (Springer, Berlin, 1996)
14. G. Harman, *Change in View: Principles of Reasoning* (Bradford Books, Cambridge, 1988)
15. L.M. Pereira, A. Saptawijaya, Counterfactuals in critical thinking with application to morality, *Model-Based Reasoning in Science and Technology. Models and Inferences: Logical, Epistemological, and Cognitive Issues*, SAPERE. Springer. Forthcoming (2016)
16. D. Davidson, *A Nice Derangement of Epitaphs*. Reprinted in Truth, Language, and History. (2005) (Oxford University Press, Oxford, 1977), pp. 89–107
17. D.M. Costa, Can creativity be formalized? Peircean reflections on the role of abduction in human intelligence, *Toward Artificial Sapience: Principles and Methods for Wise Systems* (Springer, London, 2008)
18. J. Meheus, D. Batens, A formal logic for abductive reasoning. Log. J. IGPL **14**(2), 221–236 (2006)
19. F.L. Bookstein, *Measuring and Reasoning: Numerical Inference in the Sciences* (Cambridge University Press, Cambridge, 2014)
20. G. Galilei, in *The Essential Galileo*, ed. and trans. by M.A. Finocchiaro (Hackett, Indianapolis, 2008)
21. E.O. Wilson, *Consilience: The Unity of Knowledge* (Knopf, New York, 1998)
22. P. Galison, *Image and Logic: A Material Culture of Microphysics* (University of Chicago, Chicago, 1997)
23. G. Priest, *Beyond the Limits of Thought*, 2nd edn. (Clarendon Press, Oxford, 2002)

Chapter 2
Iconicity in Peirce's Semiotics

> To him who in the love of nature holds / Communion with her visible forms, she speaks / A various language.
>
> William Cullen Bryant, *Thanatopsis*
>
> The hypothesis of the mathematician is always the conception of a system of relations. In order that they may be reasoned about mathematically, these relations must be conceived as embodied in some kind of objects; but the character of the objects, apart from the relations, is utterly immaterial. They are always made as bare, skeleton-like, or diagrammatic as possible. With mathematicians not born blind, they are always visual objects of the simplest kind, such as dots, or lines, or letters, and the like. The mathematician often passes from one mode of embodiment to another. Such a change is no change in the hypothesis but only in the diagrammatic embodiment of the hypothesis. The hypothesis itself consists in the system of relations alone.
>
> C.S. Peirce, "On the Logic of Quantity" (1895)

2.1 Introduction

We have already treated the most important aspects of Peirce's theory of abduction in the previous chapter. It will suffice here to mention just two of those earlier results: the first is that abduction is conditioned by states and state-transitions of knowledge, ignorance and experience; the second is that abductive reasoning is particularly difficult to formalize. The present chapter continues this line of investigation by focusing on Peirce's semiotic notion of iconicity. We are motivated especially by the general and informal characterization of abduction as the response to a surprising event that instigates a broadening of cognitive horizon, an opening of more or less structured possibilities out of which one or more may be selected as potential "ways forward" in explanatory understanding.

G. Caterina and R. Gangle, *Iconicity and Abduction*,
Studies in Applied Philosophy, Epistemology and Rational Ethics 29,
DOI 10.1007/978-3-319-44245-7_2

Peirce's notion of abductive reasoning may be understood to depend on iconicity in at least two respects: (α) to the extent that mathematical reasoning is taken as the paradigm of reasoning – for Peirce, all mathematical reasoning takes place through the use of diagrams and depends upon the discernment of relations internal to the diagram as well as (virtual) mappings from these relations to relations of possible tokens of similar structures; and (β) in the sense of scientific experiment, where experimental science from a pragmatist Peircean perspective is best understood in terms of diagrammatic experimentation on naturally occuring icons. Both (α) and (β) may be subsumed under the general heading of "model-based reasoning". In this way, from a Peircean perspective the relation of iconicity and abduction is coeval with the notion of model-based reasoning.

By closely examining the place of iconicity in Peirce's more general semiotic theory and in particular his notion of propositions as dicisigns, we are led to see how abductive reasoning with models depends on both iconic and indexical components. This relationship between iconic and indexical signs is instantiated both in concrete scientific investigation and abstract mathematical representation, thus indicating a common root to scientific and mathematical modelizations. The present chapter thus culminates with a formal construction based in presheaves that we call the Sheet of Indication, which is meant to serve as a quite general model for various types of diagrammatic experimentation. This construction makes explicit how iconic and indexical elements of diagrammatic reasoning are synthesized in a systematic way.

2.2 Peirce's Theory of Signs

2.2.1 Peirce's General SOI Schema

The basic unit of analysis for Peirce's doctrine of semiotics is, of course, the sign. For Peirce, a sign is itself conceived as an integral relation among three sign-components: (S) that which represents, (O) that which is represented, and (I) that in and for which the representational relation of S to O occurs. The letters are chosen in accordance with Peirce's own preferred terminology for these three components, or perhaps better, functional elements, of the sign: *sign* (at times, less intuitively but perhaps more precisely, *representamen*), *object* and *interpretant*.

Our purpose is neither to provide a comprehensive overview nor to attempt a synthetic reconstruction of Peirce's voluminous writings on semiotics. We intend primarily to resituate Peircean semiotics in light of categorical mathematics. In particular, the relational form of the sign as analyzed by Peirce is quite neatly captured by the convention of arrow-composition in category theory. Recall that one of the three axioms of category theory is that for any morphism f with domain A and codomain B and morphism g with domain B and codomain C, there is a unique morphism

h with domain A and codomain C such that h is identical to the composition of g following f. In the diagram below, $h = gf$. It is said that the diagram commutes. Note that in the juxtaposed diagram, the identical arrow-structure is evident and only the letters designating the objects are changed to conform to the S-O-I notation of Peirce's sign-components. What justifies the transference of the categorical notion of composition to the a priori unrelated Peircean notion of the triadic sign? Peirce's own formulations of the triadic sign-relation suggest that the irreducibility of the triadic relation that links sign, object and interpretant is a consequence of an internal coordination of relations between each of the three pairs of terms, such that the sign mediates a relation of interpretant to object.

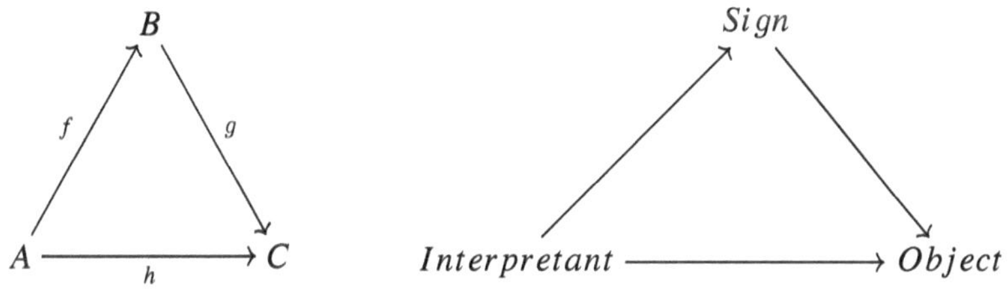

In a 1909 letter intended for William James but never sent, Peirce describes the fundamental triadic relation in the following way ([1], p. 492):

> A Sign is a Cognizable that, on the one hand, is so determined (i.e., specialized, *bestimmt*) by something *other than itself*, called its Object (or, in some cases, as if the Sign be the sentence "Cain killed Abel," in which Cain and Abel are equally Partial Objects, it may be more convenient to say that which determines the Sign is the Complexus, or Totality, of Partial Objects. And in every case the Object is accurately the Universe of which the Special Object is member, or part), while, on the other hand, it so determines some actual or potential Mind, the determination whereof I term the Interpretant created by the Sign, that Interpreting Mind is therein determined mediately by the Object.

Here, Peirce uses the language of "determination" to order the three terms of the relation: the Object determines the Sign, the Sign determines the Interpreting Mind, and the Object thus "mediately" determines the Mind. The determination itself of the Mind by the Sign is here designated by Peirce as the Interpretant, a perhaps unnecessary complication. A new draft written two weeks later restates the definition like this ([1], p. 497):

> A *Sign* is anything of either of the three Universes which, being *bestimmt* by something other than itself, called its *Object*, in its turn *bestimmt* the mind of an interpreter to a notion which I call the *Interpretant*; *and does this in such a manner* that the Interpretant is, thereby and therein, determined mediately by the Object.

Here and elsewhere, Peirce characterizes signs both in terms of the three object-like components (S, O, I) that are themselves linked by the triadic sign-relation and in terms of the types of sub-relations (here of determination or *Bestimmung*) that in fact hold among these three components.[1] On the one hand, the triadic "mediating" relation that defines the sign as such for Peirce conforms nicely to the abstract relation-composition at the heart of category theory, its basic formal element. On the other hand, however, the specific *content* of each of Peirce's terms (sign, object and interpretant) provides a way to generate more definite characterizations of such triadic mediations, that is, various *types* of signs. It is to these further determinations that we now turn.

2.2.2 Three Trichotomies and Ten Classes of Signs

Peirce distinguishes three trichotomies of sign-types (see [1], Chap. 21 as well as commentaries in [2, 3]). Intuitively, these may be thought of as successive layers of determination proceeding from more simple and general to more complex, specific and differentiated. In effect, the three trichotomies derive from a fractal-like application of each of the three components of the sign-relation to the triadic sign-relation itself. Thus signs may be categorized in terms of their type of representamen, their type of object and their type of interpretant. More precisely, the determination in each case is grounded in distinct types of relation that hold, first of all as an identity-relation of the sign itself, secondly as the type of relation between sign and object, and finally as the type of meta-relation between the relation of the interpretant to the sign on the one hand and the relation of the interpretant to the object on the other.

The First Trichotomy: Qualisign, Sinsign, Legisign

The first of the three trichotomies characterizes types of sign based on the type of being of the sign-component itself. Peirce defines these three terms as follows ([1], p. 291):

> A *Qualisign* is a quality which is a sign. It cannot actually act as a sign until it is embodied; but the embodiment has nothing to do with its character as a sign.
>
> A *Sinsign* (where the syllable *sin* is taken as meaning "being only once," as in *single, simple*, Latin *semel*, etc.) is an actual existent thing or event which is a sign. It can only be so through its qualities; so that it involves a qualisign, or rather, several qualisigns. But these qualisigns are of a peculiar kind and only form a sign through being actually embodied.
>
> A *Legisign* is a law that is a sign. This law is usually established by men. Every conventional sign is a legisign. It is not a single object, but a general type which, it has been agreed, shall be significant. Every legisign signifies through an instance of its application, which may be termed a *Replica* of it. [...] The replica is a sinsign. Thus every legisign requires sinsigns. But these are not ordinary sinsigns, such as are peculiar occurrences that are regarded as significant. Nor would the replica be significant if it were not for the law which renders it so.

[1] The clearest well-known exemplar of this tendency is found in the discussion of triadic relations in Peirce's 1903 Syllabus. See [1], pp. 289–91.

Essentially, this first trichotomy is an ontological one. It corresponds so closely as to be effectively identical with Peirce's phenomenological categories of firstness, secondness and thirdness.[2] It is not clear that by restricting these categories as being applicable only to signs, Peirce is really making a distinction that makes any difference. At the very least Peirce flirts with the idea of equating semiotics with ontology (he speaks of "a universe perfused with signs, if it is not made entirely of signs" (cited in [4], p. 104)), and at any rate we do not seem to encounter anything in actual experience that is not composed solely of constituents of these three types.

Importantly, a qualisign is defined through its independence in principle from any actual embodiment. It is in this sense that it corresponds to sheer possibility as such. A sinsign, on the other hand, only exists as such through its actual instantiation of one or more qualisigns. It thus depends upon some collection of qualisigns to be what it is. Similarly, a legisign depends upon its instantiation through particular replicas, each of which is in itself a sinsign, but whose capacity to signify as they do is grounded not in their actual, particular existences but instead their belonging in common to a general type.

The Second Trichotomy: Icon, Index, Symbol

The next trichotomy is grounded in the type of relation that holds between the sign and the object of the sign. Thus the sign-types that emerge at this level are best understood as classifications of the characteristic dyadic relation that obtains between signs and their objects, namely the general relation of signification or representation as such. This general relation of signification admits, according to Peirce, of three fundamental and exhaustive species. This triad of signifying relation-types is generated by allowing the three basic Peircean categories of firstness, secondness and thirdness (or, equivalently, the essential characters of qualisigns, sinsigns and legisigns respectively) to act in a formally differentiating way on the "raw material" of the generic sign-object signifying relation. What thereby naturally emerge are three types or species of this relation itself: one that is "qualisign-like" or a Peircean "first", one that is "sinsign-like" or a "second", and lastly one that is "legisign-like" or a "third". Peirce's terms for these three types are then, in order, icons, indices and symbols ([1], pp. 291–2):

> An *Icon* is a sign which refers to the Object that it denotes merely by virtue of characters of its own and which it possesses, just the same, whether any such Object actually exists or not. [...]
>
> An *Index* is a sign which refers to the Object that it denotes by virtue of being really affected by that Object. [...]
>
> A *Symbol* is a sign which refers to the Object that it denotes by virtue of a law, usually an association of general ideas, which operates to cause the Symbol to be interpreted as referring to that Object. [...]

---

[2] For analysis of these fundamental concepts of Peirce's philosophy, see among many similar discussions [1], pp. 4–5 and 160–1.

Thus, the specific type of relation that holds in any particular case between its sign and its object may be, in itself, qualisign-like, sinsign-like, or legisign-like. Crucially, the criterion in play here that distinguishes among icons, indices and symbols is applied not to the sign itself but to the relation between the sign and its object. In this way, the "same" triad of types is generated in a new way when lifted to the level of dyadic sign-object relations.

The Third Trichotomy: Rheme, Dicisign, Argument

Peirce's third trichotomy of signs stands with respect to the second trichotomy much as the second stands to the first. Whereas the second trichotomy lifts the three fundamental types of sign as distinguished by the first trichotomy (qualisign, sinsign, legisign) to the level of the dyadic relation between sign and object (icon, index, symbol), the third trichotomy lifts this same generic triad from the dyadic relation of sign and object to the triadic relation through which the representation of the object by the sign is in fact effective for some interpretant ([1], p. 292):

> A *Rheme* is a sign which for its Interpretant, is a sign of qualitative possibility, that is, is understood as representing such and such a kind of possible Object. [...]
>
> A *Dicent Sign* [or *Dicisign*] is a sign which, for its Interpretant, is a sign of actual existence. [...]
>
> An *Argument* is a sign which, for its Interpretant, is a sign of law.

After presenting these three definitions, Peirce immediately restates them in slightly different and perhaps somewhat more helpful terms (ibid.):

> Or we may say that a Rheme is a sign which is understood to represent its Object in its characters merely; that a Dicisign is a sign which is understood to represent its Object in respect to actual existence; and that an Argument is a sign which is understood to represent its Object in its character as sign.

Peirce concedes that "these definitions touch upon points at this time much in dispute" and adjoins "a word [...] in defense of them", discussing primarily the relation between the dicent sign and the form of propositional judgments, a proposition being "the matter upon which the act of judging is exercised", and the truth-tending orientation of any argument, the "law, in some shape, which the argument urges" (ibid.). For our purposes, the important point is the overall architectonic that structures the entire hierarchy of types: a reflexive application of the fundamental sign-triad to each of its constituent (and cumulatively determined) relations.

From Three Trichotomoies to Ten Classes

The three trichotomies combine to form ten structurally consistent classes of signs, as organized by Peirce in the following triangular table ([1], p. 296):

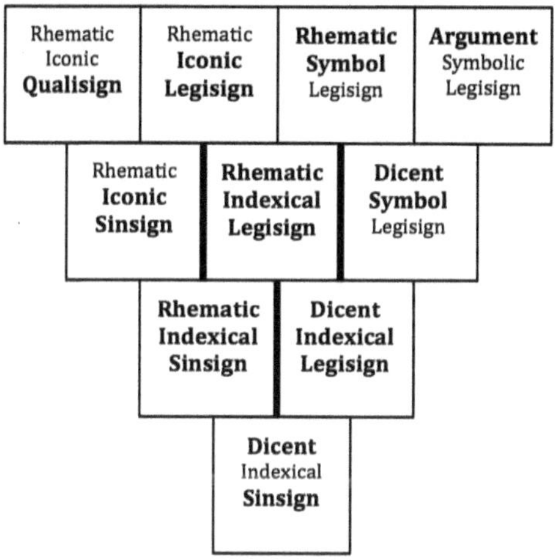

Peirce glosses this diagrammatic arrangement as follows (ibid.):

> The affinities of the ten classes are exhibited by arranging their designations in the triangular table here shown, which has heavy boundaries between adjacent squares that are appropriated to classes alike in only one respect. All other adjacent squares pertain to classes alike in two respects. Squares not adjacent pertain to classes alike in one respect only, except that each of the three squares at the vertices of the triangle pertains to a class differing in all three respects from the classes to which the squares along the opposite side of the triangle are appropriated. The lightly printed designations are superfluous.

We leave this fascinating typological puzzle to the reader's perusal without further comment.

2.3 Analyzing Iconicity

In the densely populated thicket of Peirce's semiotic theory, we are interested in the particular role of *iconicity*.

Peirce's notion of iconicity takes shape in the context of his second tripartite division of types of signs into icons, indexes and symbols. It is possible to discern three different kinds of definition of iconicity in Peirce's work, each of which corresponds to a certain point of view taken on iconic signs: as pure quality; as relational structure; and as epistemically determined as a support for abductive reasoning. This last conception will prove to be the most important one for the present argument by providing a natural way to link the semiotic notion of iconicity to the broad notion of model-based reasoning, which in turn serves as a useful framework for treating both scientific experimentation and mathematical theories of nature in a unified fashion.

The following sections outline these three distinct conceptions of iconicity and provide a progressive sequence of examples from category theory to illustrate them. The basic notions of category theory needed to follow this section are outlined in the first Appendix. See also the discussion of Peircean semiotics and the categorical expression of diagrammatic reasoning in [4], Chaps. 3–4.

2.3.1 *Iconicity as Quality*

Peirce sometimes refers to icons as signs that signify what they do independently of any question of the real existence or non-existence of their objects. In other words, icons understood in this fashion are pure possibilities or free-floating qualities. To describe a quality, say redness, as free-floating is to emphasize its essential independence from any particular conditions of instantiation, such as the redness of this fabric here under present lighting conditions, or even any instantiation at all. From a Peircean point of view, this ensures that such qualities are utterly distinct from Lockean or Humean sensations, which are concrete perceptual events with definite qualitative properties, not the pure qualitative essences expressed in and by those events. The Peircean notion of iconicity as pure quality or essential possibility is in this respect quite close to the Husserlian phenomenological concept of *eidos*. See [5] for a detailed analysis of this connection.

The essential contrast here is with indices. An index "is *existent*, in that its being does not consist in any *qualities*, but in its effects – in its actually acting and being acted on, so long as this action and suffering endures" ([6] 6.318 cited in [7], p. 81). An icon, on the contrary, at least as conceived in its purest possible state, would be a quality or collection of qualities whose being consists of those very qualities themselves understood as independent of any actual effects. Without any reference to concrete instantiations in determinate domains, icons at this level are understood as immanent systems of relations. They are in this sense "structures", but structures conceived only in terms of their intrinsic determinations, not as in the next properly structural level which treats such systems of relational determination in terms of their possible instantiations, transformations and comparative mappings.

The following figure illustrates this purely qualitative conception of iconicity within category theory as exemplified by the finite category **3**. The category **3** consists of three otherwise undetermined objects and three non-identity arrows as pictured in the diagram. By the axioms of category theory, the diagram is thus forced to commute. Identity arrows on each of the three objects are implicit, that is, present in the category but for ease of presentation not depicted in the figure. This category naturally represents a linear order on three elements (including identities) with the elements themselves left unspecified.

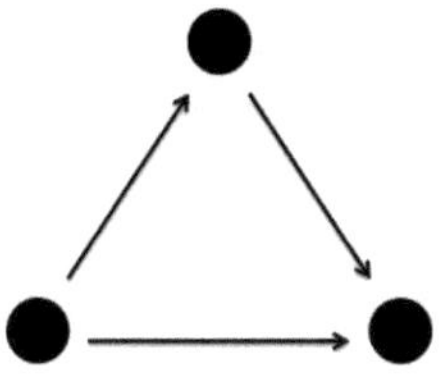

Note that this "abstract" category, that is, the kind of category whose objects are specified only as "mere dots" or structural placeholders rather than individuals of some type (such as, say, sets) with independently determined properties, provides a very natural representation of qualitative iconicity since the system of arrows and arrow-compositions that defines the category also exhausts the category's "content". There is simply nothing else to the category **3** than the linear order on three otherwise unspecified elements itself. In fact, *any* abstract category is qualitatively iconic in precisely this sense. Such categories simply *are* the systems of arrows and arrow-compositions that associate some class of otherwise undetermined objects in a determinate, systematic way.

2.3.2 Iconicity as Structure

A second more elaborated concept of iconicity understands icons in terms of the relational structure composing their parts. This notion of iconicity as structure corresponds quite closely to Peirce's own definition of icons in his second trichotomy in terms of the relation between the iconic sign and its signified object. In order for a sign and its object to share a qualitative or structural homology, it must be possible in principle to abstract the quality or structure at issue and to recognize it as the same (or relevantly similar) in the two semiotic roles.

In practice, it is difficult if not impossible to conceive of this second conception of iconicity as wholly distinct from the previous one. Instead, the conceptions of iconicity as quality and as structure should be understood as existing along a continuum with no absolutely determinate point of rupture separating the one from the other. This is largely because ordinary habits of cognition are immediately situated among complex strata of abstraction and generality. The very idea of a pure quality or a pure structure, that is to say, a structure defined in itself without any reference to its possible homologies, appears as a kind of vanishing limit, like the concept of the unrepresentable or unrepeatable. The very difference between the two is itself dependent upon the second conception. From the standpoint that recognizes quality or structure as such, it is possible to attribute one or another particular structure or quality to some object or sign, retroactively as it were. But from within the perspective that recognizes only qualities or structures in themselves, there is no available framework to distinguish and compare one with another. Each is like a universe unto itself.

To continue with the categorical example introduced above, the difference between the notions of iconicity as quality and iconicity as structure can be shown by "lifting" the previous category **3** from the level of its immanent or intrinsic relational determination to the level of the functors it supports into other categories, particularly those "concrete" categories whose objects are independently determined. The arrows in the diagram below from **3** into the categories indicated represent functors that take **3** as domain and the designated categories (sets and functions, groups and group homomorphisms, and some partial order $\leq$ defined on a given set P) as codomain.

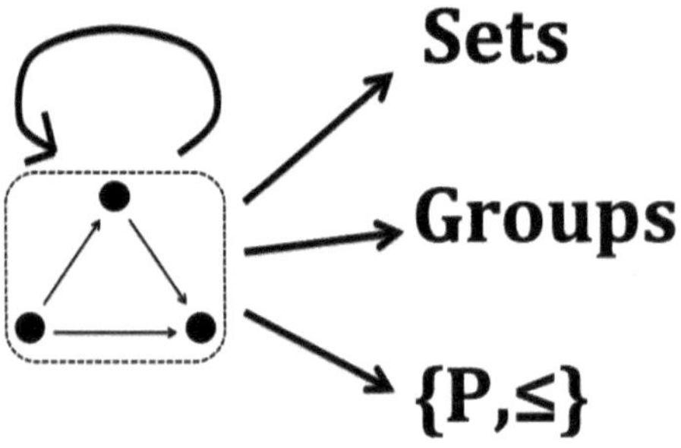

Here the functors may be understood to represent selections of the generic structure-type represented by the category **3** as instantiated in the concrete categories at issue. For example, the functor shown from **3** into **Sets** selects three sets s_1, s_2 and s_3 and three functions $f : s_1 \rightarrow s_2$, $g : s_2 \rightarrow s_3$ and $h : s_1 \rightarrow s_3$ that compose in the codomain category **Sets** according to the schematic conditions provided by the arrow-relations in the domain category **3**. That is, $gf = h$. Composition of set-functions in the "concrete" codomain mirrors composition of arrows in the "abstract" domain. Identity functions in **Sets**, like identity arrows in **3** are here implicit and unremarked upon.

Similarly, the functor represented by the arrow from **3** into **Groups** designates the selection of a composition of two group homomorphisms, and the functor from **3** into the partial order $P, \leq$ selects three elements of P, say p_1, p_2 and p_3 such that $p_1 \leq p_2 \leq p_3$. In each case, the purely relational structure immanent to the arrows and compositions of **3** is instantiated in the codomain category in terms of the specific type of relation (functions, homomorphisms and order-relation) obtaining among the objects constituting the category.

Common structure serves here as the basis for possible relation and comparison across diverse types of mathematical objects. Importantly, the "space" of possible functors defined in this way includes relatively degenerate mappings in which differences in the domain collapse to identities in the codomain. For instance, the functor from **3** into the partial order on P may take all the objects of **3** to the same element p of P. Thus some of the immanent relational structure of **3** is lost but not contravened.

Interestingly, the close relationship between this structural conception and the previous qualitative one is nicely captured by a straightforward construction in category theory. Every category C supports an identity functor 1_C that takes itself as both domain and codomain and is defined by mapping all objects and arrows of C to themselves. This identity functor expresses how the internal structure of any category, the relational dynamics of the category as an object in itself, is always also determined from a more external perspective as a functorial *relation* of that category to itself. In a similar way, every element of nature, experience and knowledge of whatever sort may, from a Peircean semiotic point of view, be understood as an iconic sign in which the sign and sign-object coincide, that is, a self-signifying icon.

2.3.3 Iconicity as Abductive Support

Finally, we turn to the third and most fully elaborated notion of iconicity as an epistemically determined relation between a sign and its object that is capable of supporting discovery, via investigation of the properties and relations of the sign, of new knowledge concerning the sign's object.

This epistemic conception of iconicity is especially important because although it remains independent of the question of the factual existence of its object, it nonetheless implies a "space" of relational instantiation in which it itself is implicated. In other words, the epistemic conception of iconicity is only intelligible if the notion of knowledge it involves is broad enough to include knowledge of "ideal" or purely structural "objects". Peirce writes, "A pure icon can convey no positive or factual information; for it affords no assurance that there is any such thing in nature" ([6] 4.447 cited in [8], p. 216). How is it possible for knowledge to be generated without any reference to "positive or factual information"? The icon instantiates, in itself, a system of relations, and this system may be investigated as such.

It could be argued that in this way every icon involves some minimal indexicality, namely the indexical relationships among its relational parts. On this view, icons would only subsist as models. In Peirce's terminology, what is relevant here is the difference and the relation between a "qualisign" and an "iconic sinsign". As Short points out, some of Peirce's 1903 writings introduce the technical term "hypoicon" to represent the specific way that an actual existent (a sinsign) serves as the "substratum supporting an icon" ([8], p. 216; see also [9]). This terminology is useful for present purposes, as it suggests the existential "basis" for iconic properties or values that "float over" it. This picture is obviously similar to the construction that we employed at the beginning of chapter one. This picture is at once closely related to and distinct from the traditional formalization in first-order logic between "individuals" as elements of a domain over which quantifiers are allowed to range and the properties that adhere to these. It is distinct insofar as the icons or properties that are linked to particular individuals are not regulated by logical determinations such as binary truth-values. Instead, they are treated as "occasions" for concrete experimentation.

In the figure below, a functor F from the category **3** into $P, \leq$ is investigated by means of functors from the one-object, one-(identity)arrow category **1** into **3**.

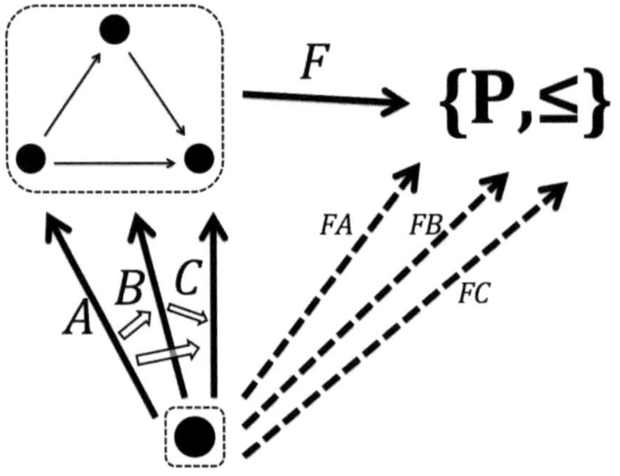

The three functors A, B, C: $\mathbf{1}\longrightarrow\mathbf{3}$ select the initial, medial and terminal objects respectively of **3** and are related to one another via the natural transformations depicted by the hollow arrows near the base of the diagram (which compose in the obvious way). These three functors from **1** into **3** induce three functors into the partial order, namely FA, FB, FC: $\mathbf{1}\longrightarrow P, \leq$ via composition with F. Here, the structural mapping provided by the functor F is taken as given, and this mapping itself is investigated by probing **3**, the domain of F, directly with functors from a third category and ascertaining the results of such probing in the codomain of F as entailed by composition with F. The structure of the codomain of the functor is "known" mediately via direct functorial relations with the functor's domain.

The three ways of understanding Peirce's concept of iconicity are in this way both instantiated and clarified by examples drawn from the mathematics of category theory. In short, to recapitulate:

1. An icon conceived in itself is an immanently determined relational structure.
2. An icon conceived in relation to its object(s) is a structure determined by its possible translations and transformations.
3. An icon conceived in relation to its interpretant is an instantiated structure that may be investigated abductively so as to generate potential knowledge about some objective domain that it models.

2.4 Iconicity in Peirce's Theory of Dicisigns

A crucial role is played in Peirce's system of semiotic classification by the second term of his third trichotomy, the proposition, or *dicisign*. For Peirce, a dicisign is a sign-type defined by its interpretant's orientation towards the signified object of the sign as being one of actual existence. Thus a dicisign is understood as a way of treating a sign such that the object of the sign is conceived to exist in reality. In this respect, dicisigns are essential elements of abductive reasoning to the extent that abductive inferences are concrete "existential" responses to genuine epistemic problems.

The role of iconicity in Peirce's theory of propositions is especially striking. Stjernfelt's study *Natural Propositions* [7] synthesizes Peirce's widely scattered writings on this notion and extracts a definite theory of propositions that aligns very closely with Peirce's more general theory of semiotics as well as his philosophy of science. What Peirce calls a dicisign is what the tradition after Peirce will usually call a proposition, although it seems clear after the work of Stjernfelt that the Peircean dicisign is in fact a generalization of the more common notion of proposition, that Peirce's concept envelops not only ordinary linguistic propositions but also certain types of images and other sign-forms.

A dicisign is a type of sign that coordinates an iconic and an indexical function. Peirce's unusual view with respect to the analysis of propositional semiotics amounts to a coordination of syntactical with semantical structure. Yet the coordination at issue is not what might be expected: a homology between the relational composition of the relevant parts of the representing domain and the relations among the semantic elements in the domain of what is represented. Such would be the structural homology view of propositional meaning advocated, for instance, by the early Wittgenstein of the *Tractatus*. Instead of such a view, Peirce proposes a theory of propositions in which the indexical component is just as important as the iconic one in generating propositional sense.

Recall that Peirce introduces dicisigns (his own *terminus technicus* for what are now commonly called propositions) in the context of his third trichotomy of signs. Peirce understands what he calls a "rhema" as a fragmentary part of a logical discourse. It is closely related to what in a Scholastic vocabulary would be called a "term" with the important difference that a rhema is intrinsically fragmentary in character. It is best expressed by a notation such as "—is mortal" rather than the mere word or term "mortal". It includes the copula (the relation or connection to a subject) as an intrinsic yet underdetermined part of itself (see [1], pp. 308–9). Peirce's rhema is thus quite similar in form and function to what Frege develops as the notion of an "unsaturated" concept.

The following excerpt establishes the basic framework for Peirce's theory of propositions ([1], pp. 310–11):

> It may be asked what is the nature of the sign which joins "Socrates" to "—is wise" so as to make the proposition "Socrates is wise." I reply that it is an index. But, it may be objected, an index has for its object a thing *hic et nunc*, while a sign is not such a thing. This is true, if under "thing" we include singular events, which are the only things that are strictly *hic et nunc*. But it is not the two signs "Socrates" and "wise" that are connected, but the *replicas* of them used in the sentence. We do not say that "—is wise", as a general sign, is connected specially with Socrates, but only that it is so as here used. The two replicas of the words "Socrates" and "wise" are *hic et nunc*, and their junction is a part of their occurrence *hic et nunc*. They form a pair of reacting things which the index of connection denotes in their present reaction, and not in a general way; although it is possible to generalize the mode of this reaction like any other. There will be no objection to a generalization which shall call the mark of junction a *copula*, provided it be recognized that, in itself, it is not general, but is an *index*. No other kind of sign would answer the purpose; no general verb "is" can express

> it. For something would have to bring the general sense of that general verb down to the case in hand. An index alone can do this. But how is this index to signify the connection? In the only way in which any index can ever signify anything; by involving an *icon*. The sign itself is a connection. [...] It is, thus, clear that the vital spark of every proposition, the peculiar propositional element of the proposition, is an indexical proposition; an index involving an icon. The rhema, say "—love—", has blanks which suggest filling; and a concrete actual connection of a subject with each blank monstrates the connection of ideas.

The basic method, then, of abductive reasoning with respect to propositions may be characterized in terms of the controlled variation of the iconic and indexical components of dicisigns. This coordination of abductive inference with the Peircean conception of propositions has the virtue of providing theoretical continuity between the analysis of abduction with respect to propositions on the one hand and with respect to diagrams on the other. Indeed from this perspective propositions appear as a special case of diagrams.

What generally makes a diagram successful as an instrument of abduction is the epistemic advantage gained by transferring investigation into some object of inquiry from a relatively inaccessible or experimentally intransigent domain to one that is more readily manipulated. This strategy is typically that of "model-building" or "model-based reasoning".

In *Natural Propositions*, Stjernfelt introduces a distinction between "operational iconicity" and "optimal iconicity". Operational iconicity is essentially Stjernfelt's reformulation of the epistemological determination of iconicity discussed above. This conception gives rise to an "operational criterion" for identifying iconic signs: "icons [are] the only sign type able to provide information" ([7], pp. 207–8). By this, Stjernfelt means that only iconic signs carry information about their objects when investigated independently of their signifying relation itself. Clearly, an index, say, a weather vane, carries information about its object, the direction of the wind. The point at stake here, however, concerns the relative externality or internality of the information thus carried. The "internal" representation of information within some medium is essentially diagrammatic.

In distinction from this "operational" determination of icons, which as Stjernfelt emphasizes tends to extend the notion of diagram "to such a degree that the common-sense notion of diagrams vanishes in the haze and seems to constitute only a small subset of the new, enlarged category" (ibid., p. 211), a stronger criterion, "optimal iconicity", is introduced. Sternfelt's own example is taken from Peirce's EG_β level of Existential Graphs and its "line of identity" that represents the existential quantifier in that logical system in an iconic way.

The problem here is that Stjernfelt attempts to ground the distinction between operational and optimal iconicity in an "absolute", or decontextualized determination of degrees of iconicity. But this cannot be done, or at least very significant obstacles present themselves. A very suggestive symptom of the issues at stake is provided by Stjernfelt's own characterization of the EG_β example he cites. Sternfelt writes (ibid., p. 218):

> The substitution of selectives for the line of identity is less iconic because it requires the symbolic convention of identifying different line segments by means of attached identical

> symbols. The line of identity, on the other hand, is immediately an icon of identity because it makes use of the continuity of the line that, so to speak, just stretches the identity represented by the spot – and that is, at the same time, a natural iconical representation of a general concept [...].

The problem may be highlighted simply by drawing attention to the phrase "so to speak" in the citation just given.

Nonetheless the most important point about Peirce's interpretation of propositions as dicisigns is that it makes essential use of a particular theory of relations. Specifically, it depends upon the claim that relations imply infinitely iterable repetitions that nonetheless remain semantically inert in the sense that they do not enter into the interpretation of the meaning of the relation as such. The idea is illustrated by the following figure:

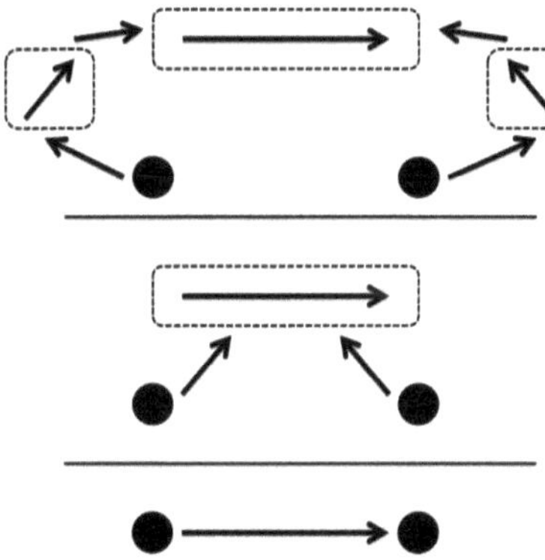

Peirce's theory effectively internalizes the synthetic operation of the relation (its active "joining" of its respective relata in its own characteristic manner) to the relation's own objecthood. Without further analysis, we merely note here that this semiotic and ultimately metaphysical conception of relations is well-captured by the the role of identity-arrows in category theory.

2.5 Representation and Hypostatic Abstraction

> So [human beings] maintained it as certain that the judgments of the Gods far surpass man's grasp. This alone, of course, would have caused the truth to be hidden from the human race to eternity, if Mathematics, which is concerned not with ends, but only with the essences and properties of figures, had not shown men another standard of truth.
>
> Spinoza, *Ethica* I

Dicisigns are the semiotic bread and butter of representations of fact. They are signs taken as describing what some real state of affairs is actually like. For Peirce, it is the iconic aspect of the dicisign that enables the semiotic relation of sign to reality to hold *for* some interpretant. In particular, in scientific knowing it is the iconic character of whatever discourse is employed for expressing scientific statements that on this view subtends the potential ability of those statements adequately to represent the real. Thus iconicity grounds the very possibility of scientific representation. When the discourse at stake for representing natural reality is essentially mathematical, as for instance throughout physics, it is the iconic character of mathematics that makes its representational function with respect to nature possible. For this reason, only a genuine understanding of iconicity in mathematics will support a robust scientific epistemology and, in particular, a clear grasp of abductive mathematical reasoning in natural science.

It is helpful here to look internally to mathematics, as the standard notion of representation in mathematics is itself an iconic one. Mathematics of course employs various modes of representation in its notational conventions, including for instance multiple sorts of variables (essentially indexical signs) and a variety of notations for representing formulas, operations and equations (essentially symbols). But within this very large space of mathematical representations, there is the relatively narrow conception of "representation" in the distinctly mathematical sense.

A standard mathematical expedient is to represent an abstract structure (typically algebraic or order-theoretic) within the more "concrete" mathematical domain of sets. Of course, sets are by no means concrete in any ordinary or absolute sense. They are mathematically concrete, however, precisely insofar as the theory that governs the overall structure of their common domain is quite well-studied and well-understood, and moreover to the extent that this domain may be conceived as a common space for determining the relationships among effectively all other types of mathematical objects and structures (as in the monumental twentieth-century synthesis arranged by Bourbaki). The "structure" of sets may be determined in various ways, most commonly by axiomatizations of the primitive epsilon-relation (as in Zermelo–Fraenkel or Gödel–Bernays) or through categorical characterization of the category **Sets** of sets and functions.

In many cases, a simple way to translate abstract, purely relational structures into sets is to convert each relevant abstract term into the set of the relations that hold between it and all the other relevant abstract terms at issue. The relations among the abstract terms are then carried over into perfectly isomorphic relations among the newly formed concrete objects just defined.

How? Expressed mathematically, it consists of the establishment of a kind of mapping between two domains that is guaranteed to preserve the relevant structure of one inside another. In category theory, this type of mapping is typically modeled by a functor. Illuminating examples of this are found in Cayley's representation theorem and categorically in the Yoneda lemma. What goes by the name representation theory in contemporary mathematics is a rich elaboration and extension of the sort of method used by Cayley in demonstrating his important result. Rather than representing algebraic structures in sets, representation theory typically translates abstract algebraic structures into linear transformations of vector spaces, thus allowing relatively "concrete" operations on matrices to capture the relevant "abstract" structures at issue.

The very existence within mathematics of Cayley's theorem and such branches of research as representation theory naturally raises the question: What relation, if any, holds between this type of representation as determined internally to mathematics and the general notion of representation at work in the world outside of mathematics that structures, for instance, Peircean semiotics? More specifically, we may now distinguish three broad problems of "representation", each inhabiting its own self-determined space of inquiry:

1. What is representation and how does it function within mathematics across various mathematical domains?
2. What is representation and how does it function between various areas of mathematics and various worldly domains, particularly as employed by the sciences?
3. What is representation and how does it function in a general way in the natural world itself (including of course that most dear part of the natural world that includes us human beings, our cultural products, sciences and so forth)?

We have ordered these questions in a perhaps counter-intuitive way, from the seemingly specific and regional (internal to mathematics) to the apparently most general (semiotics as the general theory of natural and cultural representation via signs), precisely so as to suggest that the otherwise obvious subsumption of mathematical knowledge and truth as a subclass of all kinds of knowledge and truth might be fundamentally mistaken. The undeniable and, at least from perspectives like that of Wigner, astonishing success of mathematics in scientific modeling of the world might in fact reflect the essential and fundamental role of mathematical thinking in *all* forms of abductive reasoning rather than its status as a special case. Perhaps, then, it is possible within mathematics to develop a theory of relations that would include the scientific usages of mathematics and the natural processes of semiotics under its proper purview. What is wanted then is a mathematical theory of representation general enough and flexible enough to accommodate both purely mathematical examples – Cayley's theorem, representations of algebraic structures as linear transformations of vector spaces, etc. – as well as the relatively impure cases of scientific measurement and experimentation of natural processes themselves.

2.5.1 Hypostatic Abstraction

Peirce's semiotic notion of *hypostatic abstraction* provides this needed bridge between mathematical method and the role of representation in mathematics on the one hand and abductive reasoning in natural science on the other. Whereas pure mathematical reasoning, on Peirce's view, considers nothing other than the universe of sheer iconic possibility, scientific inquiry is essentially propositional, that is, it coordinates iconic and indexical elements.

The notion of diagrammatic abduction already complicates this all-too-easy distinction. Peirce maintains that all mathematical reasoning is intrinsically diagrammatic. It thus requires reasoning not only about pure icons, but reasoning concerning pure icons on the basis of *instantiated* icons, that is, diagrams that involve both iconic and indexical dimensions. The mathematical reasoner must be able to *experiment* on the icons that serve as the theoretical objects of her inquiry, and such experimentation requires that the relations pertaining to the icons at stake be manipulable and thus indexically available for concrete interaction. Thus the semiotic "secondness" of actual, reactive and resistant existence is not so easily subtracted from the essence of mathematical reasoning. If it does not enter directly into the ultimate theoretical *objects* of such reasoning, it nonetheless remains an ineluctable aspect of the reasoning process as such.

In a somewhat different fashion, Peirce also keenly understood the importance of the unique and unusual role played by the concept of existence in the sphere of mathematical reasoning, even at the level of mathematical objects as such. The specifically mathematical, or iconic, notion of existence is in contrast to the standard idea of existential secondness as passive, event-like reaction. Instead, in a purely iconic register the objects or "entities" of mathematics are thoroughly constructive, as befits their status as qualisigns or pure possibilities. In this regard, Peirce was at pains to emphasize the role of *hypostatic abstraction* in mathematical reasoning.

The method of hypostatic abstraction is typically abductive in character. It posits or constructs an "existential" hypothesis that tends to play an explanatory or elucidating role in its given cognitional context. This abductive role of hypostatic abstraction links mathematical reasoning directly to scientific inquiry, where the "existential" frame is not in general one of pure iconic possibility but rather of the reactive secondness of causes and effects linked in sequences of real events. In natural science, explanation is typically causal and concerns relations between actual entities. For Peirce, however, the mathematically constructive method of hypostatic abstraction is nonetheless the primary means for conducting scientific inquiry into such relations and the patterns they exhibit.

Peirce's own preferred example, to which he returns in a variety of texts and contexts, is that of the "dormitive virtue" attributed to opium by the Scholastically trained doctor in Molière's satirical comedy. For Peirce, the doctor's seemingly vacuous pretension is in fact a theoretical technique of tremendous subtlely and power. Let us analyze this canonical instance of "empty" reasoning more closely:

1. The reasoning begins with the recognition of a causal relation between the substance opium and the property of somnolence:

   $$\text{opium} \rightarrow \text{sleep}$$

   This step is already in its own way a kind of hypothesis. As the Humean tradition rightly insists, causal relations as such are not observable; they are posited. Nonetheless, one starts here with what is effectively an induction sufficiently provocative of surprise or wonder.
2. The surprise at "seeing" the causal connection between opium and sleep raises the question of the *explanation* of this causal relation. What stands in relation to

this relation as such, in such manner that it would become "a matter of course"? Informally:

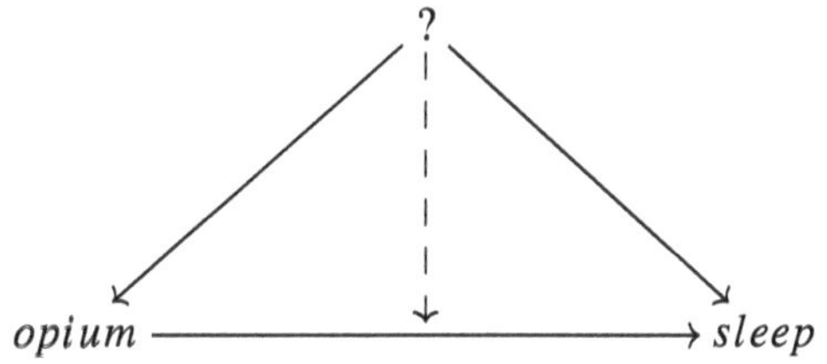

This question opens an enormous space, unrestricted in principle, of possible explanations. The second stage of the abductive procedure is thus introduced: how might this too-large space be appropriately reduced to something more manageable, that is, less questionable and more positively suggestive?

3. What immediately suggests itself is the theoretically "minimal" solution that would simply repeat the causal arrow posited in step 1 as a new "object", that is, an "external" cause that is at the same time a property attributable to the term "opium".

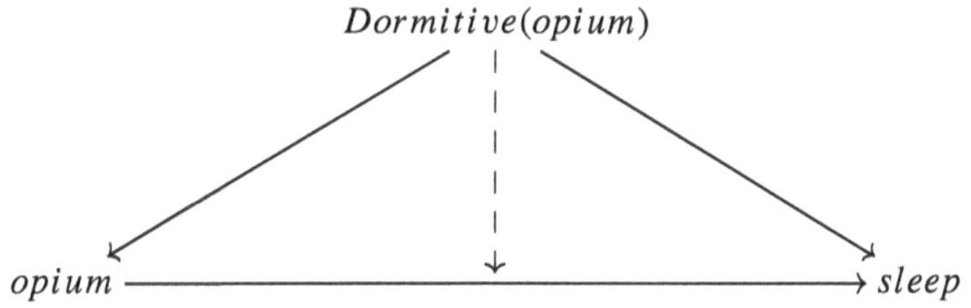

This procedure induces an implicit argument-scheme very much in accord with the general Peircean model of abductive inference as "affirming the consequent":

$$\frac{\begin{array}{l}\text{opium} \longrightarrow \text{sleep}\\ \text{Dormitive(opium)} \rightarrow (\text{opium} \longrightarrow \text{sleep})\end{array}}{\text{Dormitive(opium)}}$$

4. From a slightly different point of view, the previous abductive step may be understood not as positing "dormitive virtue" as an external cause, but rather as *internalizing* the causal arrow (the relation) in step 1 to its first term "opium" (its "initial" relatum).

No matter whether the abductive process of hypostatic abstraction as just sketched is interpreted more "externally" (as in step 3) or in a more "internal" manner (as suggested in step 4), the hypothesis thereby formulated that opium possesses dormitive virtue is not merely a restatement of the original datum that opium causes somnolence. What the abductive-conclusion-as-hypothesis provides is a newly enriched space of reasoning that supports new kinds of questions ("Is dormitive virtue somehow like the liveliness virtue of caffeine, even if opposite in effect?") and new forms of inquiry ("Perhaps looking at opium through a microscope will reveal the mechanism of dormitive virtue").

2.5.2 *Iconicity and Axiomatics*

An analogue of hypostatic abstraction appears in its own way in modern mathematics. In Hilbert's notion of axiomatic method, any system of axioms that does not imply a contradiction may be understood to be satisfied by purely ideal "objects" that are determined only by the axioms themselves. Such an axiom system is required only to be consistent. The mathematical "formalism" that results rejects any limitation of mathematics by natural ontology. The mere fact that a model of an arbitrary axiom system cannot be given in terms of some domain that actually exists is irrelevant to the question of the mathematical truth of that axiom system. More strongly, the axiom system itself guarantees the objects it discerns.

Importantly, however, the elements of the syntax in an axiomatic system are intrinsically variable. They are indifferent as to everything but their syntactical character, which is nonetheless dependent upon pre-existing systems of determinable content. The very possibility of designating a regime of variables with a characterization such as "a, b, c ..." requires the prior grouping of the lower-case Roman alphabet as well as the cognitive organization of a "Roman-letter type". Rodin, in [10], examines in detail the history of axiomatic method from its earliest emergence in Greek geometry through its initial modern treatment in Hilbert and his followers to its contemporary employment in nearly all branches of mathematics as well as many fields of science.

Rodin follows Lawvere (see especially [11, 12]) in noting a deep connection between the categorical approach to logic and axiomatics on the one hand and the Hegelian notion on the other of an objective logic that would remain immanent to the real dialectic of the "thing itself" understood as an epistemic process in which objective and subjective aspects would ultimately cohere. From the in-itself of purely objective being and the for-itself of simple subjective interiority, a process at once real and epistemically oriented would culminate in the in-and-for-itself of subjective and objective reconciliation. See also [13] in this regard.

In particular, Rodin translates the distinction made by Hegel himself in [14] between a merely "subjective logic" such as that proposed by Kantian idealism and Hegel's own innovative notion of an "objective logic" of dialectical reality into the difference between the traditional use of logical deduction as an *external* framework for examining the consequences of mathematical axioms and the structure, particular to topos theory, of an *internal* logic. Among other things, Rodin shows how the "bootstrapping" approach to topos theoretical axiomatization in McClarty [15] solves in a strictly mathematical register the philosophical problem of immanent logical grounding that appears in Hegel as the problem of overcoming the division of idealism and realism.

Rodin is at pains to show that what goes by the name "formalism" in philosophy of mathematics and is commonly attributed to Hilbert as its originator is in fact rooted in a misreading of the father of modern axiomatic method. Furthermore, in this regard the logicist program that followed Frege and Russell in aiming to ground mathematics in logic also represents a misreading and misapplication of logic in the context of axiomatics as understood by Hilbert. Rodin points out that just as the geometric axioms of Euclid do not simply express logical relationships among truth-claims but in fact underwrite the possibility of concrete *constructions* in the ideal geometrical plane, the formal axiomatic method of Hilbert does not represent the application of independently conceivable logical operations to syntactical data. Rather, the syntax-semantics relation (the core of model theory) is already internal to formal logic as such.

He cites Hilbert's well-known 1927 paper on "Foundations of Mathematics" (cited in [10], p. 61):

> No more than any other science can mathematics be founded by logic alone; rather, as a condition for the use of logical inferences and the performance of logical operations, something must already be given to us in our faculty of representation, certain extralogical concrete objects that are intuitively present as immediate experience prior to all thought. If logical inference is to be reliable, it must be possible to survey these objects completely in all their parts, and the fact that they occur, that they differ from one another, and that they follow each other, or are concatenated, is immediately given intuitively, together with the objects, as something that neither can be reduced to anything else nor requires reduction. This is the basic philosophical position that I regard as requisite for mathematics and, in general, for all scientific thinking, understanding, and communication. And in mathematics, in particular, what we consider is the concrete signs themselves, whose shape, according to the conception we have adopted, is immediately clear and recognizable. This is the very least that must be presupposed; no scientific thinker can dispense with it, and therefore everyone must maintain it, consciously or not.

When Hilbert claims here that "as a condition for the use of logical inferences and the performance of logical operations, something must already be given to us in our faculty of representation", we may prescind from the Kantian language of "faculties" and attend only to the basic point at stake, namely that logical reasoning (even in its most formal, non-psychologistic sense) operates necessarily on some sort of "material", some sort of non-logical data. Even and especially when this data is purely symbolic, that is, formal, it must nonetheless be instantiated in a system of possible relations (a generic "space") and that space itself must be in its own way *given as evident and available*. The point, philosophically speaking, is an Aristotelian one: no forms without matter; *hylē* and *morphē* are mutually implicative concepts. Yet, from a purely mathematical perspective it may *also* be compatible with a sort of Platonism. The "matter" of mathematical reasoning may itself be purely ideal (eidetic), as in the standard treatment of ZFC-consistent sets as a kind of raw material for building mathematical structures. Interestingly, Hilbert indicates a trio of conditions that appear to govern the very possibility of any such logically operable "material":

1. Existence: "that they occur".
2. Difference: "that they differ from one another".
3. Synthetic relation: "that they follow each other, or are concatenated".

The mathematics of category theory may be understood in this way as a distillation of Hilbert's program, one that deploys it *internally* to mathematics itself.

2.6 Iconicity and Diagrammatic Abduction

At the heart of abductive reasoning there is a sort of reconstruction problem. Abduction is indeed characterized by the gluing of "local", context-dependent information, into a synthetic, coherent global organization of possibilities. In mathematics, the notion of *sheaf* plays a similar role. Since we believe that the role of sheaves is fundamental in the understanding of the notion of iconicity within abduction, in this section we present a short introduction to presheaves and sheaves.

Following an example presented in [16], consider the role of a judge whose job is to evaluate the artistic quality of painted hard-boiled eggs. Clearly, given the shape of the object being evaluated, the judge faces a subtle challenge since she – at any given time – only has a partial view on the whole object (she can only see half of the egg from any given angle).

Obviously, the judge reasons under the assumption that, by rotating the eggs, she will have enough partial points of view that, all taken together, will give her enough information to reconstruct the global picture and therefore make an informed decision on the artistic quality of the painted egg.

In mathematics we can find a very general version of this process: first, the notion of presheaf formalizes the notion of "locality", whereas the notion of sheaf corresponds to the conditions that need to be fulfilled to glue together local information in order to get the global picture.

Definition 1 Let X be a topological space. A *presheaf* of abelian groups on X, say F, is an object F which associates an abelian group $F(U)$ to every open subset $U \subset X$ such that

1. for any inclusion of open sets $V \subset U$ we have a group homomorphism

$$\rho_{VU} : F(U) \longrightarrow F(V)$$

2. for every open U, ρ_{UU} is the identity map on U
3. $W \subset V \subset U \Rightarrow \rho_{WU} = \rho_{WV} \circ \rho_{VU}$

The group $F(U)$ is called the group of *sections* of U and the map ρ_{VU} is called the restriction of $F(U)$ onto $F(V)$.

Example 2 Real (topological) manifolds are topological spaces that, locally, look like the Euclidean space of a fixed dimension. Topologists often study the presheaf of *smooth functions* over a manifold. In other words, given an open set U, $F(U)$ is the ring (hence an abelian group) of all smooth functions over U. In this case the effect of the restriction map ρ_{VU} on a section $f \in V$, is just the actual restriction function $f|_V$.

We remark that, instead of abelian groups, we could have used any category and the definition of presheaf would still be the same. The existence of a unique gluing procedure characterizes the sheaf condition.

Definition 2 A *sheaf* F is a presheaf that satisfies the following conditions. Let $\{U_i\}_{i \in I}$ be a collection of open sets in X and let U be the union of $\{U_i\}_{i \in I}$.

1. (Existence of gluing) Whenever, for every $i \in I, s_i \in F(U_i)$ and for every $i, j \in I, s_i|_{U_i \cap U_j} = s_j|_{U_i \cap U_j}$ there exists a section $s \in F(U)$ with $s|_{U_i} = s|_{U_j}$ for all $i \in I$.
2. (Uniquiness of gluing) Let $s, t \in F(U)$ such that $s|_{U_i} = t|_{U_i}$ for all $i \in I$. Then $s = t$.

In the subsections below we develop a formal framework aligned with presheaves to characterize the intrinsic syntax-semantics relation of iconicity as it appears in diagrammatic representations quite generally. The clearest motivating intuition will come from considering ready-to-hand, concrete instances. As an example, consider a computerized touchscreen, perhaps as found on a contemporary handheld tablet device. On the screen are several icons, windows, tabs and so forth. The user is able to select from among the various objects available on the screen by touching the screen itself with his or her finger and adopting various simple conventions (tapping twice to select an object, etc.). In particular, he or she may be able to select arbitrary collections of objects on the screen (perhaps by holding the SHIFT key down while selecting individual objects in succession).

Imagine now a generalization of this function of selecting parts by indicating. Think of a sheet that may be inscribed with any image, picture or complex figure and imagine a weblike overlay that may be set to the precise contours of useful, selectable parts of this image, picture or figure enabling reference to some external domain in some communicative context. For instance, someone might select a certain area of a projected map and say, "The suspect is hiding somewhere in here."

What we call the Sheet of Indication is meant to function something like this idealized map or touchscreen. It is intended to function as a formal semantics that is applicable to a wide variety of diagrams and, as we will see in the next chapter, to the logical diagrams of Peirce's Existential Graphs (EG) in particular.

2.6.1 *Introducing the Sheet of Indication*

What is the Sheet of Indication (SI)? It is any sheet inscribed with some image or diagram, certain determinate parts of which may be indicated in a given communicative context. Thus we have two basic components:

- the inscribed sheet itself
- a system of indication that depends upon a conventional and rule-governed semiotic relation (*symbolic*, in Peirce's terms) on the basis of which it becomes possible to indicate (*indexically*) certain determinate parts of the inscription.

We require that the parts that may be indicated or selected in this fashion determine amongst themselves a partial order (that is, a relation that is reflexive, transitive and anti-symmetric) and that there be a privileged part, called *terminal*, that is greater than or equal to all the parts according to the order. Together, these conditions ensure a certain synthetic and tractable coherence among the parts.

We then associate to each selectable part of the diagram – that is, to each element of the partial order – a set of possible values in some determinate domain. In this way, each selectable part functions somewhat like a variable in a formal syntax. Next we define models for the diagram as a whole in terms of the coordination of sets of value-assignments to the various parts of the diagram that respect in an appropriate way the order among these parts, depending on the specific features and intended uses of the diagram at issue.

In order to motivate the formalization of this semantics, we begin with a pair of informal examples that will allow us to clarify the main features of the construction. Our first example is a simple Venn diagram consisting of two overlapping circles.

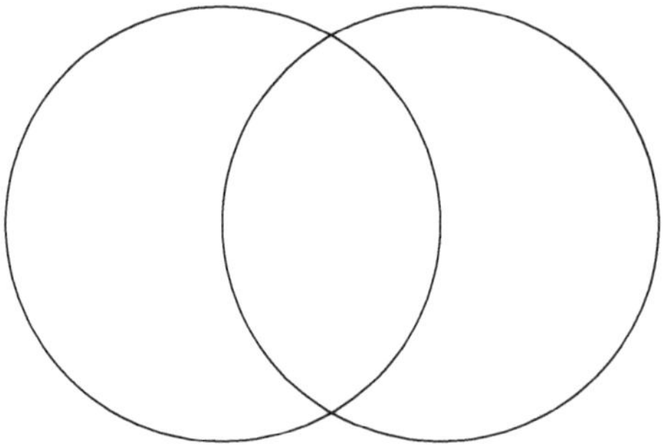

In order to conceive this diagram as a Sheet of Indication, we must first come to an agreement concerning which parts of the diagram may be distinctly indicated. We must then ensure that these parts are ordered amongst themselves by some relation that is reflexive, transitive and antisymmetric. Also, there must be some part at the "top" of this order that stands in this relation to all the other parts. In the present case, we may stipulate that any bounded area of the diagram and any collection of such areas may be uniquely indicated (for instance, by touching one or more areas of the diagram simultaneously). The parts that may be selected in this fashion are then themselves naturally ordered by the relation of inclusion: collections of bounded areas include the areas they collect.

The following "exploded" diagram makes explicit what such parts are and how the order-relation works in this particular case. All the selectable parts are individually represented by collections of areas marked by dots, and an arrow from one such part to another represents that the part at the "tail" of the arrow is "below" the part at the "head" of the arrow in the partial order. For simplicity's sake, we have omitted arrows that may be inferred from the reflexivity and transitivity of this order-relation.

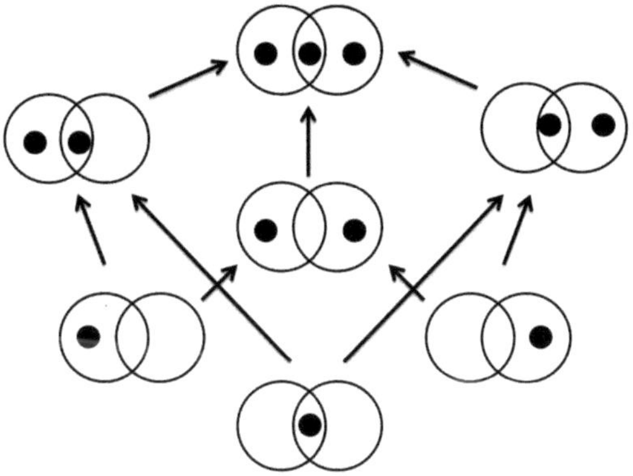

Our concern is to formulate a diagrammatic semantics. How might we actually use the original two-circle diagram in some given context to represent values in the relevant domain of reference? Let us assume we wish to classify teachers working at a particular school. We might naturally assign to each of the parts of the diagram some collection of these teachers. It is clear that according to the standard interpretation of Venn diagrams, we would want the inclusion relations among those subsets to correspond exactly to the arrows (both drawn and implied) in the exploded diagram above, that is, to the partial order defined over the selectable parts of the diagram.

More generally, in using this diagram and above all in employing it in a regular way across various contexts, we would want to introduce certain fixed general conditions on possible assignments of subsets of some set S to a particular part of the diagram that would depend on the assignments to certain other parts. A complete model for the diagram in any given context would then consist of some definite assignment of values to every part that respects these conditions. The information contained in such a model could then be "read off" the largest part, that is, the area containing all the other areas. Some particular distribution of math and science teachers, for instance, could be represented by associating the math teachers with the left circle and the science teachers with the right circle, with those teachers who teach both math and science associated with the lozenge-shaped intersection of the two circles. All of this information would be contained and properly differentiated by the complete determination of the most inclusive area of the diagram, in this case the diagram as a whole.

A second example shows how the same basic idea might be applied to a different sort of diagram with a different mode of semantic determination. In the figure below, we have a diagram representing a sports playoff bracket. Here, the rectangles are meant to represent teams and the dots to represent games. Individual dots and rectangles are the parts that may be selected, and the order defined over these parts is given by the arrows shown. Once again the order relation is reflexive and transitive, so arrows from each part to itself and arrows corresponding to every possible path of drawn arrows are implied. Whereas in the previous example all the selectable parts of the diagram were of the same kind, namely, bounded areas, in this diagram we have two different types of part corresponding to two different types of assignment: dots corresponding to games and rectangular boxes corresponding to teams.

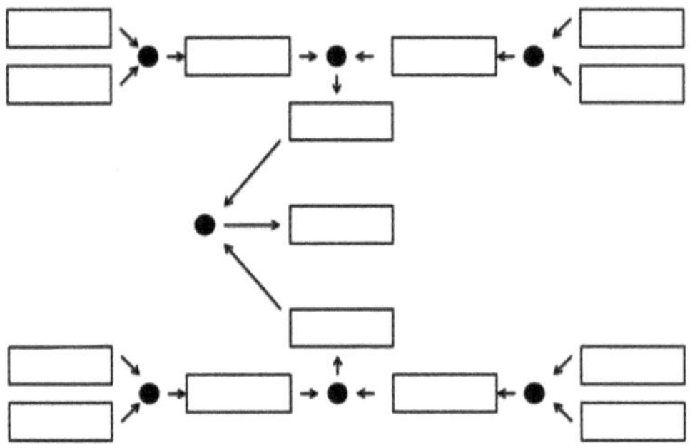

Given some collection of teams, say those constituting a particular conference division, we would want to build a possible model for the diagram by assigning one of the teams to each rectangle and a pair of teams to each dot. Obviously, to maintain the coherence of the semantics, we would want in addition to impose a set of conditions on such assignments. For instance, we would want to ensure that no team can play against itself, that the game assigned to a given dot will consist of precisely the two teams assigned to the rectangles immediately preceding it, that the winner of any game assigned to a dot will be assigned to the rectangle immediately succeeding it, and so on. A complete determination of the diagram will then be given by the assignment of a unique champion team to the rectangle in the center, together with an assignment to all the parts "below" this part of a team or game consistent with the specified conditions.

2.6.2 *Formal Definition of the Sheet of Indication*

We now define a Sheet of Indication formally.

The *Sheet of Indication*, or SI, is a diagram consisting of:

1. a finite set P of *parts* together with a partial order $(P, \leq)$ endowed with a terminal object p^* (that is, $\forall q \in P\ q \leq p^*$)
2. a finite set of *types* T and a function $t : P \longrightarrow T$ that assigns a type to every part.

A semantics for the SI is then given by specifying functions from *lower sets* of the parts of the diagram into some *domain of interpretation*. The domain of interpretation is any non-empty set D, and a lower set of any part $p \in P$ is defined as the set of all parts less than or equal to p, that is, $\{q \in P | q \leq p\}$. Thus the semantics ranges over functions that we call *models at* p, or m_p, for all $p \in P$:

$$m_p : \{q \in P | q \leq p\} \longrightarrow D$$

We designate the set of all such specified models at a given part p with the notation μ_p.

For any $m_p \in \mu_p$, we define the *value at* q for any $q \leq p$ to be $m_p(q)$, that is, the value assigned to the part q in the model m_p. We denote the value at q in the model m_p by $val_{m_p}(q)$, although we usually drop the index when the model m_p is clear from the context.

Constraints on the functions m_p are specified by a set C of *conditions*, that is, statements that stipulate values (or ranges of values) in D that must or may not be assigned in models m_p to parts $q \leq p$, depending on the order-relation $\leq$ and the type-function t. Any function not explicitly required or prohibited by a condition is implicitly allowed. The only restriction on sets of conditions themselves is that they must never permit some model at p, $m_p \in \mu_p$, while also prohibiting m_p as restricted to the lower set at any $q \leq p$ to be a model at q, $m_q \in \mu_q$.

Let us see how our previous examples fit this formal framework. Parts and types in both cases are obvious. Equally clear are the domains D of the interpretation. What really specifies a particular SI are the conditions that constrain the values in D that the models may assign to distinct parts and typed parts. In the case of the Venn diagram, for instance, there is only one type (all the parts are collections of bounded areas), so only the order-relation $\leq$ is relevant to the conditions on the models. Clearly, we want to ensure that the relations between values at each part in all models correspond to the appropriate set inclusion and intersection operators. This may be guaranteed by a set of three conditions that apply to values in all models at all parts:

- $q \leq p \Rightarrow val(q) \subseteq val(p)$;
- $q \nleq p$ and $p \nleq q \Rightarrow val(q) \bigcap val(p) = \emptyset$;
- $\exists q\ (q \leq p$ and $q \neq p) \Rightarrow val(p) = \bigcup_{r \leq p, r \neq p} val(r)$.

In a similar fashion, we can recover the relevant conditions on models for the playoff bracket diagram. The main difference there is that its semantics involves two different types of parts: games and teams. Presumably with respect to these types at least two of the conditions we would demand of any model would be:

- $t(p) = game \Rightarrow val(p) = \bigcup_{r \leq p, r \neq p, \forall q | r \leq q \leq p \Rightarrow r = q} val(r)$
- $t(p) = game \Rightarrow \exists x, y | x \in val(p)$ and $y \in val(p)$ and $x \neq y$

Thus if there are no two-element sets in the domain D of interpretation, there will be no models for the diagram and the resulting semantics may be said to be *empty*. Assuming an appropriate domain of interpretation, however, here too an adequate set of conditions would function not only for this graph in particular, but for any similarly formed graph (such as a similarly structured playoff bracket with a greater number of rounds). In this way, a set of conditions C functions roughly as a classifier for diagrams that are to be interpreted semantically "in the same way".

The most important thing to notice in both examples is the way that the models in the semantics may be conceived roughly as being built up layer by layer along the contours of the partial order $\leq$ from the "bottom" to the "top." What that means formally is that for any model at $p \in P$ and for any $q \leq p, q \in P$ there is always a model at q which is the model at p restricted to q (the two functions agree on all values). Because the collection of models over each part form a set, we then have a naturally defined function $f_{p,q}$ from the set of models at p, μ_p, to the set of models at q, μ_q for any $q \leq p$. This function takes each model in μ_p to the corresponding model in μ_q that is simply its restriction to q. In addition, such functions compose in a natural way. That is, given any three parts $r \leq q \leq p$ and the three functions $f_{q,r}$ and $f_{p,q}$, we have that $f_{p,q} \circ f_{q,r} = f_{p,r}$. Defined in this way, the models over P have the structure of a *presheaf*.

In [17] the distinctively philosophical importance of presheaves (and the sheaves they induce) is stressed insofar as the most philosophically interesting aspects of advanced contemporary mathematics are both generated by and best represented in terms of synthetic local/global relations across various mathematical domains. Because presheaves and sheaves lend themselves naturally to modeling such relations, they are especially useful in studying structures that vary over the elements of other structures and the way such systems may be layered or "glued" to one another (see [18] and especially [19] for details and references). In [20] this approach via sheaves and presheaves is applied specifically to the philosophy of Peirce and the distinctive understanding of continuity that underlies the "Synechism" of his later philosophy. The present diagrammatic semantics aligns nicely with this general approach to both a wide range of issues in contemporary philosophy of mathematics and logic and to Peirce's philosophy and semiotics in particular. Anticipating our analysis in chapter three, related research that interprets Peirce's EG_α in terms of different category theoretic constructions is found in [21, 22].

For those readers who may be unfamiliar with presheaves, the following diagram may help make the notion more vivid:

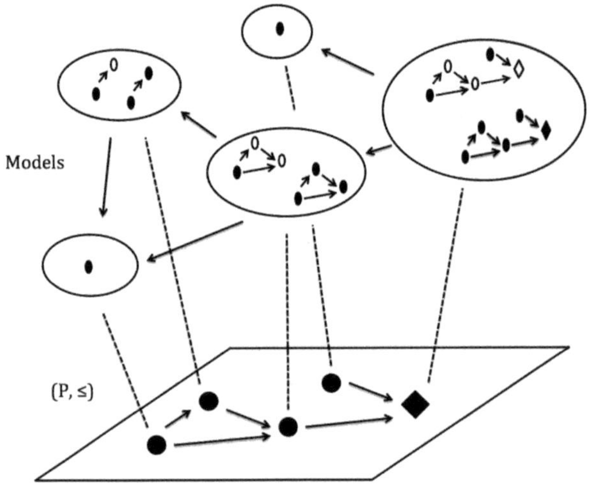

The presheaf structure in the above diagram is evident by noticing how the natural restriction maps between the models "reverses" the order given in the base space. The reader may him or herself trace how each of the two models in the diagram at the terminal part represented by the diamond on the sheet corresponds naturally to a model at each of the other parts that "looks exactly the same" when restricted to the "piece" of the model at and below that part. Tracing these restrictions simply involves following the arrows connecting the sets of models one to another.

In general, there may be models at a part p for some sheets of indication that are *not* included in any model at a part $q > p$. Thus there could have been two distinct models in the set over the "least" part of the diagram above (the southwesternmost part) and the restriction maps would have remained exactly the same and the overall structure would still be a pre-sheaf. Nonetheless in practice many if not most interpretations of diagrams induce restriction maps between models that also happen to be surjective. What is impossible in any case because of the required pre-sheaf structure is that there be any model m at a part p such that there is a part $q \leq p$ and there is no model at q that takes exactly the same value as m at each part $r \leq q$. This entails that a restriction map from the set μ_p of models at p to the set μ_q of models at q for any $q \leq p$ is always well-defined.

We thus have a general formal framework (the SI) that quite naturally represents the syntactical and semantical dynamics of representation with respect to the mereological (part-whole) relations of a wide variety of different types of diagrams. The interaction of iconic and indexical components of such diagrams is reflected in the identity conditions and structural relations of the parts themselves.

2.7 Conclusion

Let us review the steps we've taken in this chapter:

- We sketched the main contours of Peirce's general semiotic theory and focused on the notion of iconicity in particular.
- We distinguished three levels of iconicity corresponding to quality, structure and abductive support. Each of these levels found a natural representation in the terms of category theory, thus suggesting a link between semiotic iconicity and categorical mathematics.
- We examined Stjernfelt's analysis of Peirce's doctrine of dicisigns, which establishes a coordination of iconic and indexical semiotic elements as the primary mechanism whereby propositions become capable of referring to the world.
- We considered Peirce's notion of hypostatic abstraction relative to the general problem of propositional and mathematical representation on the one hand and that of abduction on the other.
- We developed the formal theory of Sheets of Indication to model systems of diagrammatic experimentation in a general way.

Thus, by examining the concept of iconicity in detail we have been led to reformulate abductive reasoning as a diagrammatic process involving both iconic and indexical components, thus bridging the gap between purely "ideal" iconic qualities and the "real" structures where such icons are instantiated and experimented upon. This intrication of iconic and indexical elements of abductive reasoning may be analyzed in its own right with the use of tools drawn from the mathematics of category theory. In the following chapter, we show how the general syntax-semantics relation of iconic diagrams applies in the particular case of Peirce's logical notation of Existential Graphs by recasting Peirce's graphs in terms of the Sheet of Indication formalization defined above.

References

1. C.S. Peirce. Ed. Peirce Edition Project. (1998). The Essential Peirce: Selected Philosophical Writings, vol. 2 (1893–1913) (Indiana University Press, Bloomington and Indianapolis)
2. P. Farias, J. Queiroz, On diagrams for Peirce's 10, 28, and 66 classes of signs. Semiotica **147**(1/4), 165–184 (2003)
3. R.W. Burch, Peirce's 10, 28, and 66 Sign-Types: The Simplest Mathematics. Semiotica **184**, 93–98 (2011)
4. R. Gangle, *Diagrammatic Immanence: Category Theory and Philosophy* (Edinburgh University Press, Edinbergh, 2016)
5. F. Stjernfelt, *Diagrammatology: An Investigation on the Borderlines of Phenomenology, Ontology, and Semiotics* (Springer, Berlin, 2007)
6. C.S. Peirce, *Collected Papers of Charles Sanders Peirce*, vols 1–6 (Belknap Press, Cambridge, 1931–1935)
7. F. Stjernfelt, *Natural Propositions: The Actuality of Peirce's Doctrine of Dicisigns* (Docent Press, Boston, 2014)
8. T.L. Short, *Peirce's Theory of Signs* (Cambridge University Press, Cambridge, 2007)
9. P. Farias, J. Queiroz, Images, Diagrams, and Metaphors: Hypoicons in the Context of Peirce's Sixty-Six-Fold Classification of Signs. Semiotica **162**, 287–307 (2006)
10. A. Rodin, *Axiomatic Method and Category Theory* (Springer, Heidelberg, 2014)
11. W. Lawvere, Foundations and Applications: axiomatization and education. Bull. Symb. Log. **9**(2), 213–224 (2003)
12. W. Lawvere, Functorial semantics of algebraic theories and some algebraic problems in the context of functorial semantics of algebraic theories. Repr. Theory Appl. Categ. **5**, 1–121 (2004)
13. B. Mélès, Pratique mathmatique et lectures de Hegel, de Jean Cavaillès à William Lawvere. Philosophia Scientiae **16**(1), 153–182 (2012)
14. G.W.F. Hegel, *Science of Logic* (Prometheus Books, Amherst, 1991)
15. C. McClarty, *Elementary Categories* (Elementary Toposes, Oxford, 1992)
16. T. Lovering, Sheaf Theory, Preprint, https://tlovering.files.wordpress.com/2011/04/sheaftheory.pdf
17. F. Zalamea, *Synthetic Philosophy of Contemporary Mathematics* (Urbanomic, 2012)
18. S. Mac Lane, *Categories for the Working Mathematician* (Springer, Heidelberg, 1998)
19. M. Reyes, G. Reyes, H. Zolfaghari, *Generic Figures and Their Glueings* (Polimetrica, Milan, 2004)
20. F. Zalamea, *Peirce's Logic of Continuity: A Conceptual and Mathematical Approach* (Docent Press, Chestnut Hill, 2012)
21. G. Brady, T.H. Trimble, A categorical interpretation of C.S. Peirce's propositional logic Alpha. J. Pure Appl. Algebr. **49**, 213–239 (2000)
22. G. Caterina, R. Gangle, Iconicity and abduction: a categorical approach to creative hypothesis-formation in Peirce's existential graphs. Log. J. IGPL **21**(6), 1028–1043 (2013)

Chapter 3
Categorical Iconicity in Peirce's Existential Graphs

> [I]n a sense logic is a special case of geometry.
> W. Lawvere [1]

3.1 Introduction

The notion of iconicity developed in the previous chapter might appear open to the objection that the use of any such conception of iconic "structure" to model abduction must err either on the side of being too formal and thus insensitive to the concrete, existential details of the situation at hand, or that of being too concretely figural or "intuitive" and thus inappropriate for the rigorous purposes of logical analysis. In fact, we aim to show that the particular power of iconic relations to support robust abductive modeling rests precisely in their capacity to reduce this Scylla-and-Charybdis problem of dry formalism versus muddy intuition to the status of a false dichotomy. Especially when modeled and analyzed via the categorical mathematics of presheaves and sheaves, iconicity as such provides a conceptual framework that is capable of varying smoothly and continuously (and tractably) from the most concrete, material domains to the most formal and abstract.

In this chapter we provide a concrete example of a category that integrates both "external" iconic structure through universal properties instantiated in a categorical manner (initial and terminal objects, products and coproducts, and exponentiation) and "internal" iconicity characterizing the category's objects as such. The former relationships are expressed by categorical relations ("arrows" or morphisms) between objects, whereas the latter are expressed by topological properties instantiated in the objects themselves, which are in fact iconic logical diagrams. The category will thus be shown to be iconic, in the sense developed in chapter two, at two different levels.

G. Caterina and R. Gangle, *Iconicity and Abduction*,
Studies in Applied Philosophy, Epistemology and Rational Ethics 29,
DOI 10.1007/978-3-319-44245-7_3

Iconicity appears both as a structure immanent to the system as a whole as well as being a property of each of the system's objects taken individually. Most importantly, the relationships between these two distinct levels of iconicity can be shown to be easily tracked by smooth and elegant diagrammatic techniques.

The system that we will examine is that of the Existential Graphs developed by Charles Peirce. This diagrammatic system of logic (and, more simply, the fragment of it – EG_α – that faithfully represents classical propositional logic) stands here as a powerful illustration of the logical and abductive power of iconicity. The chapter's sections are ordered according to the following sequence:

- Sketching some of the overall motivations in Peirce for developing an iconic logic.
- Introducing the main features of Peirce's system of EG_α and proving a simple but important theorem.
- Demonstrating how individual EG_α graphs may be expressed as diagrammatic Sheets of Indication based on presheaves (our main formal construction from the previous chapter) in a way that exhibits the iconic alignment of their topological and logical properties.
- Examining the diagrammatic relations *between* individual graphs as determined by Peirce's deductive transformation rules.
- Constructing the category $\mathcal{G}$ incorporating all the logical relations expressed by the transformation rules in a holistic manner that can be tracked by categorical universal mapping properties corresponding to Peircean diagrammatic proofs.
- Expressing abductive inference in this context in terms of categorical adjunction (adjoint functors).

3.2 Motivating Iconic Logic

In his well known early essay "Some Consequences of Four Incapacities," C.S. Peirce critiques the dominant Cartesian assumptions of modern philosophy and lays out the key principles for the alternative that Peirce himself proposes, above all the principle of an intrinsic relation between a semiotic conception of thinking and a pragmatic notion of truth as the formation and testing of hypotheses. This link between signs (triadic relations) and hypotheses (abductive inferences) will continue to prove crucial throughout Peirce's later oeuvre.

One main point Peirce makes with respect to Descartes concerns the Cartesian method of doubt. In opposition to Descartes's radical conception of systematic skepticism which is meant to clear the way for an unshakeable ground of absolute certainty, Peirce ([2], p. 40) maintains that

> [w]e cannot begin with complete doubt. We must begin with all the prejudices which we actually have when we enter upon the study of philosophy. These prejudices are not to be dispelled by a maxim, for they are things which it does not occur to us can be questioned.

In contrast to Descartes, Peirce insists that doubt and certainty, as well as all other modes of thinking, remain essentially situated within a definite context. No one thinks purely "in general". Thought is always a specific response to a more or less definite problem that arises in a determinate context. Thus, Peirce advises (ibid.),

> Let us not pretend to doubt in philosophy what we do not doubt in our hearts.

Peirce's notion of determinately situated inquiry has profound implications for the study and theorization of abductive inference. Abduction is the process whereby in the course of inquiry, in the wake of some unexplained phenomenon or "surprising event", it becomes necessary to produce a hypothesis (one as "good" as possible) that would account for the phenomenon in question. One of the most interesting and controversial questions surrounding abduction, as indicated in chapter one, continues to be whether or to what extent this process of abductive inference may itself be formalized. As already discussed, this question has been posed very explicitly by several authors and in different contexts: Aliseda [3] and Walton [4] focus on the logical formalism of abduction, whereas Magnani [5] emphasizes the role of abduction in cognitive science and its interdisciplinary connections. It is now generally accepted that while abduction can be modeled within specific frameworks, there is no single adequate model of abductive inference. More importantly, it appears there can be none, since the creative aspect of abduction remains intrinsically open to unforeseeable possibilities.

One of the reasons for the deadlock in the construction of formal models of abductive inference has been the unquestioned assumption of a clear distinction between syntax and semantics in formal logic. This distinction is so fully ingrained in contemporary logic that it would seem almost to pass beyond questioning. Yet so long as syntax and semantics are held to be independent, the work of formal logicians on abduction often repeats the Cartesian error that Peirce was at pains to criticize. A formal syntax is meant to abstract away from any determinate "content", that is, any particular semantics, much as Cartesian doubt is meant to generalize skepticism across all possible beliefs. As soon the problem of creative hypothesis-formation is posed, one must range across all possible hypotheses in a similar fashion. The problem, to be sure, is that in such cases there is no well-defined, pre-statable set of *all* beliefs or *all* hypotheses (real possibility exceeds and outstrips actuality and representability in principle). The construction of a hypothesis thus at times requires a creative act, a "guess" that is not merely the selection of one among a set of pre-given candidates. Any problem for which an abductive inference is truly necessary is necessarily an actual and thus determinately particular problem. There is no "problem of abduction" in general; there are only particular problems that emerge within specific trains of concrete inquiry. Such concreteness precludes the radical bracketing of semantics.

If the formalization of abductive inference is to be capable of tracking the situatedness of inquiry, it will be advantageous for the above reasons to work within a framework in which syntax and semantics are not held entirely distinct. How might such a framework be structured? A logical formalism could be developed that would be at once a syntax and a semantics for that syntax, in other words, that would possess the semiotic character of, precisely, *iconicity*.

The line of thought just sketched may suggest at least part of the motivation for Peirce's own development of an iconic logical formalism in his system of Existential Graphs (EG). This three-tier system (*alpha*, *beta* and *gamma*) corresponding to propositional, first-order predicative, and higher-order and modal logics respectively was considered by Peirce himself to be his greatest achievement and continues to generate scholarly interest and study today. Our concern in what follows is to draw out the consequences of positioning the problem of the formalization of abduction within the context of iconic logical notation and its concomitant blurring of the distinction between syntax and semantics. In particular, we will focus on Peirce's EG_α and the problem of abductive amplification within the binarism of propositional logic, in effect the introduction of an intermediate truth-value into a binary system. We will first show how the iconic nature of the representation of logical relations in EG_α allows for the distinguishing of part-whole Gestalts that correspond to "shapes" of possible inference. Different levels of visual inspection of the graphs correspond to finer and coarser scales of logical analysis. Then in a second step, we will show how these "shapes" themselves may be ordered naturally through the mathematics of partial orders. The category-theoretic treatment of partial orders allows these iconic relations themselves to be represented iconically. This suggests a broad and highly general research platform for the investigation of abductive inference, one that is capable of integrating previous formal work into a unified context and opening up unforeseen possibilities through its iconic nature.

3.3 An Iconic Model of Propositional Logic

It is often convenient in philosophy, as in mathematics, to generalize a problem and then treat the original problem as a specification of the more general one. This is essentially the method we intend to follow here with respect to Peirce's notion of the Sheet of Assertion and its fundamental role in EG's diagrammatic syntax. How is a diagrammatic (i.e. iconic) logical syntax possible in general? We propose to address this question within the limits of Peirce's EG_α by treating Peirce's EG Sheet of Assertion as well as the graphs that may appear upon it as instances of the Sheet of Indication as formalized previously in chapter two.

The system constituted by the EG_α graphs and their relations to one another both represents and instantiates certain regular logical properties. The iconicity that links the levels of representation and instantiation is broadly topological in nature. This means that the syntactical content of the graphs depends only upon properties related to the connectedness of their parts. Topologically invariant transformations of the graphs correspond to syntactically equivalent expressions [6]. An interesting comparison may be made between Peirce's approach in the alpha level of EG and Tarski's analysis in [7], which is discussed in [8] although without reference to the work of Peirce.

3.3.1 The Sheet of Assertion

The Sheet of Assertion (SA) is a key element of EG. In a sense, the entire system of EG, including all three levels of alpha, beta and gamma, develops naturally out of the very notion of the SA as the material or screen for an iconic topological syntax. The SA represents the ground of truth for a given universe of discourse. It serves as the iconic representation of the fundamental presuppositions that enable any particular discourse to function as a discourse oriented toward truth. As Roberts puts it, "What does SA express? Whatever is taken for granted at the outset to be true of the universe of discourse. The sheet thus functions as a kind of all-purpose axiom" ([9], p. 32).

The Sheet of Assertion is a strange entity, at once material and ideal: a visual screen and a diagrammatic substrate. In Peirce's parlance, a graph is scribed on the SA by a graphist ([9], pp. 31–33). The SA stands in this way as a condition for a semiotic relation between the graphist's regulated activity of construction and experimentation on certain diagrams and the logically significant formal transformations of the diagrams themselves. With respect to the inner structure of these diagrams (the relationships among their component parts), the SA itself remains curiously inert. It is in effect nothing more than a blank slate for possible inscriptions. Yet this role as an inert condition is by no means trivial. Perhaps the primary advantage of Peirce's diagrammatic logic vis-a-vis the various now-traditional algebraic systems of symbolic strings is to be found here, in the thematization of the material/ideal conditions of possibility for the syntactical manipulations involved in the logical system at issue and the inclusion of these conditions within the system itself as a distinct element of the syntax. The SA is both an enabling condition and a definite component part of the system EG.

3.3.2 The Graphical Syntax of EG_α

The Sheet of Assertion supports the iconic syntax of EG by providing the ground for the inscription of *characters*, corresponding to propositions, and *cuts*, that is, closed curves distinguishing topologically connected regions of the SA. A cut distinguishes the area comprising its interior from the area in which the cut is scribed. Since the canonical logical interpretation of a cut is that of the negation of the conjunctive content of its interior, cuts function iconically in the sense that they topologically separate their contents from their surrounding area just as the operation of logical negation works as a "cutting-away" of some determinate semantic content from a given universe of reference. The metaphoricity suggested by the quotation marks around "cutting-away" is of the essence here. Such metaphoricity (closely aligned with Peircean iconicity) is given a logically rigorous basis through the system of EG itself. In this way, EG is both an iconic logic and (at least potentially) a logic of iconicity. We may begin to unfold this possibility by observing some of the most fundamental features of the graphs. (For comprehensive introductions to EG, see [9–12]; for more focused studies of their semantics, see [13, 14]). First of all, it is important to note that the nesting of cuts gives rise to a natural partitioning of all the regions of the SA into two classes, *even* and *odd* as shown below.

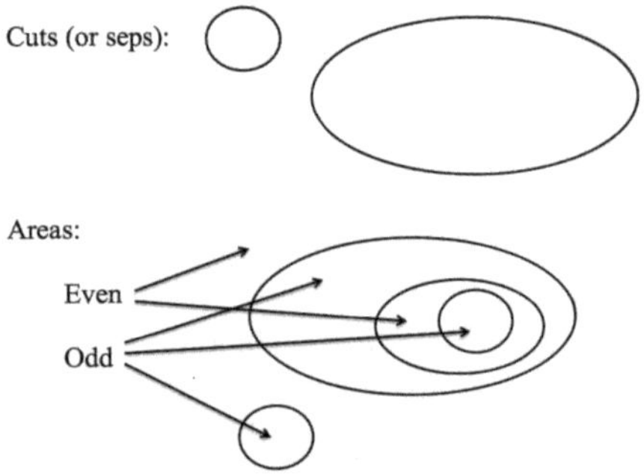

With respect to the naturally ordered nesting of areas on the SA, logical propositions are represented by symbolic tokens (conventionally, letters of the alphabet but more generally any distinctive written symbol that may be recognized and reproduced). We call any symbol of this type a *character*.

More precisely, we have the following essential components of EG_α.

1. The blank sheet (called the Sheet of Assertion or SA) is both the site on which graphs are *scribed* and is itself a graph (called the empty graph).
2. A *character* is any reproducible image that is not a *cut* (see below) scribed on part of the SA.
3. Characters may be enclosed, along with a local area surrounding them (a neighborhood on the SA that may or may not include other characters or figures), by a closed curve called a *cut*. It is convenient to construct these generally as ovals or circles. These cuts may not intersect characters, nor may they intersect one another. They may, however, be nested with any number of characters and cuts scribed in the areas or enclosures they distinguish.

Well-formed alpha graphs may then be defined according to the following recursive construction:

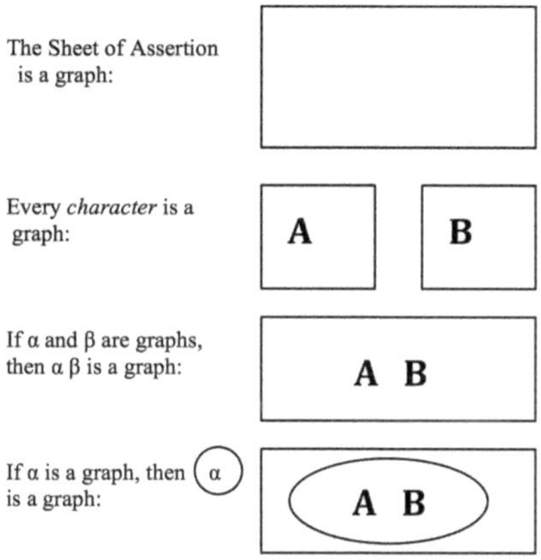

The logical properties of any given graph are then determined by the part/whole relations that structure each of its areas as well as the order of the nesting of the areas themselves. In addition, the system of EG_α as a whole is determined by the ways that graphs are permitted to be transformed into different graphs according to definite rules. Peirce's alpha transformation rules prescribe derivations of graphs from other graphs in such manner that truth is preserved under the standard interpretation of cuts as negating their conjunctive contents. In other words, they guarantee that if a graph G corresponds to a true assertion then the assertion corresponding to any graph G' that may be constructed from G via repeated application of the transformation rules will also necessarily be true. Peirce's alpha rules have been shown to be both sound and complete with respect to classical propositional logic [12]. We summarize the transformation rules here (for details see [9]):

- Write Odd: Any graph may be scribed on an odd area.
- Erase Even: Any graph may be erased from an even area.
- Iterate: Any graph may be copied onto the same area on which it is scribed or any area contained within the area on which it is scribed (at whatever depth of cuts), except on any area that is itself part of the graph to be iterated.
- Deiterate: Any graph that could have been the result of an iteration (as immediately above) may be erased.
- Double cut: Two nested cuts with nothing between them may be scribed or erased from any area around any graph on that area.

These rules are illustrated heuristically by the examples below:

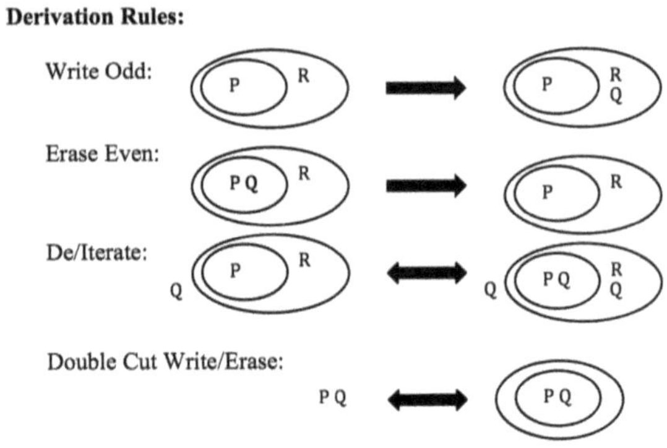

Since most of the activity involved in scribing and manipulating the graphs (technically *subgraphs*) concerns the cuts and the graphs inscribed on their areas, it is easy to forget the crucial role played by the SA in making the entire system of EG possible. It is the SA that serves as the condition of possibility for any graphical activity whatsoever. It has an intuitively synthetic function in the sense that all of its parts are continuous in the Peircean sense of continuity (hence the natural interpretation of the logical conjunction of two graphs scribed on the same area). At the same time, this continuity of the SA equally undergirds the dual notion of discontinuity (via cuts) which is naturally interpreted as logical negation. See [15] for a more detailed exposition of this point.

Another equally important feature of the SA is that it undergirds the concrete practices of reasoning that employ the graphs. When EG graphs are actually used in determinate contexts, they are generally treated as diagrams with both iconic and particularly indexical features (see [16] in particular on beta-graphs). Reasoners using EG will frequently point to, scribble on, or talk about regions, points or figures that may in general be conceived as determinate parts of the SA. For example, in an actual EG dialogical situation one speaker might very well point to the two parts of the graph below as indicated by the two arrows and say to her interlocutor, "Look, one thing this graph says is that unless this proposition here (arrow 1) is true, something on this area over here (arrow 2) must be false."

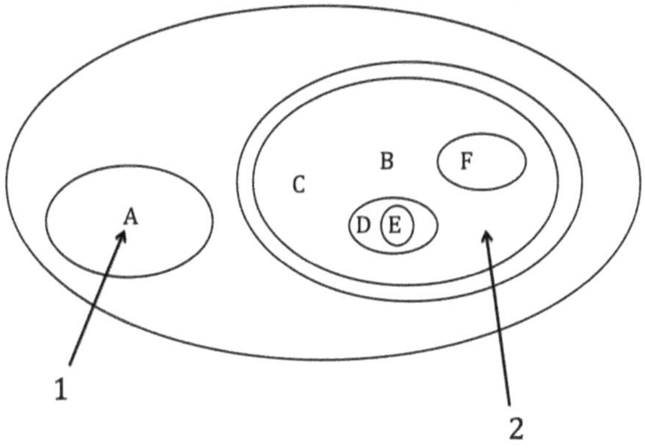

Assuming it is reasonable, how is such talk possible? Inspection of the graph in conjunction with an understanding of the EG_α conventions and some elementary logical reasoning will indeed confirm that the speaker is essentially right in what she says. However, it is difficult to translate precisely what is logically correct about her statement with respect to the diagram itself (and not merely with respect to the purely logical form that it shares with, say, a corresponding sentence expressed according to the conventions of classical propositional logic). The speaker's statement is not only more or less technically correct, but is also diagrammatically *a propos* once we recognize it to be in fact identifying a straightforward logical disjunction ($\phi \vee \neg\theta$ where θ is itself a conjunction of four statements) implicit in the form of the graph itself.

Bringing to light the relations between iconic and indexical features of the graph in the above example opens a perspective on EG and the role of the SA that situates the logical structures of the graphs in intrinsic relation to Peirce's more general semiotic categories (icon, index, symbol). Is it possible to recast such a perspective in a formal semantics? In order to help understand and also to investigate further such features of EG, we introduced in chapter two the notion of a Sheet of Indication, or SI. The semantic tool of the SI was intended to help us interpret and evaluate the ways EG is and might be used in a manner that is rigorous and yet remains sensitive to the specifically diagrammatic and hence symbolic, indexical and iconic features of EG.

The correspondence between EG_α and propositional calculus (with seps interpreted as the *negation* of their contents and the inscription of multiple graphs on the same area as their logical *conjunction*) has been exhaustively investigated by Zeman [12] and more recently in a much broader context by Brady and Trimble [17]. The iconic quality of Peirce's calculus, however, should not be reduced to the formalism of logic, since it does possess, as shown by several authors, including the knot theorist Louis Kauffmann [6], a truly topological nature. The derivation rules themselves may be characterized in a combinatorial-topological nature, in a such way that the dual nature of a graph as equally syntactical and semantical emerges naturally. The following result provides one example of this, showing how the derivation rules may be applied locally (in a topological sense) in such a way that local transformations can be embedded in larger structures.

Theorem 1 *Let G' be a graph which can be derived from a graph G. If the graph G is inscribed in an evenly enclosed area of another graph $\hat{G}$, then the graph $\hat{G}'$, obtained by replacing, in $\hat{G}$, G with G', is derivable from $\hat{G}$.*

Proof If G' is derivable from G, then there is a sequence of graphs

$$G = G_0, G_1, \ldots, G_n = G'$$

where G_i is obtained via a single application of one of the above rules from G_{i-1}. Each element of any such sequence can be lifted to a correspondent element in the sequence of graphs

$$\hat{G} = \hat{G}_0, \hat{G}_1, \ldots, \hat{G}_n = \hat{G}'$$

where by $\hat{G}_i$ we denote the graph $\hat{G}$ with G in it replaced by the correspondent G_i

For each step in the sequence, say from the ith to the $(i+1)^{th}$ either the *parity* (i.e. the evenness or oddness of the enclosures of a graph) matters or it does not in determining if $\hat{G}_{i+1}$ can be derived from $\hat{G}_i$ in agreement with the transformation rules. If it does not (De/Iterate, Double Cut), then $\hat{G}_{i+1}$ can be trivially derived from $\hat{G}_i$. If it does matter (Write Odd, Erase Even), then again $\hat{G}_{i+1}$ can be derived from $\hat{G}_i$. For otherwise we should conclude that G was enclosed in an oddly enclosed area (because only then would the parity of the relevant enclosures in $\hat{G}_i$ be different in G_i) which contradicts our assumption. *QED*

This theorem offers one instance of how certain logical properties of the graphs (syntactical relations) may be expressed as topological transformations (dynamical semantics). More generally, the topological syntax embedded in the transformation rules makes EG_α formally equivalent to classical propositional logic when characters are taken to represent distinct propositions [12]. In the next section we aim at embedding the graphs in an even more general and formally rigorous framework in which any such intrinsic relation between syntax and semantics in the graphs may be modeled and investigated.

3.4 EG_α Graphs as Sheets of Indication

We now show that every EG_α graph may be redescribed as a Sheet of Indication such that models over the sheet correspond exactly to logically coherent valuations. Two primary criteria govern this redescription: it should be general enough to apply to any well-formed EG_α graph; and it should be capable of tracking the ways such graphs are actually used (or usable) diagrammatically.

Given a Sheet of Assertion inscribed with an alpha-graph G, we assume that a diagram-user is able to indicate any of the areas, cuts and characters of the graph.

We start by declaring that given two parts $p \neq q$, q is an *immediate predecessor* of p, denoted by $q \leq_\star p$, if and only if $(q \leq r \leq p$ and $r \neq q) \Rightarrow r = p$, for all r. Next, we define:

1. the set P of parts: all areas, cuts and characters of the graph G are elements of P
2. the partial order $\leq$ defined for all $p \in P$: cuts and characters are $\leq$ the area on which they are scribed; areas are $\leq$ any cuts in which they are contained; all relations are transitive; $p \leq p$ for all parts p
3. the set of types $T = \{A, C, H_1, H_2, \ldots, H_n\}$ where A stands for areas, C for cuts, and the $H_j s$ are all the character-types in question, for instance elements of the set $\{A, B, \ldots\}$;
4. the function $t : P \longrightarrow T$ takes each part p of the graph to its type;
5. the domain of the interpretation $D = \{\bigcirc, \square\}$, where $\square$ represents the empty sheet and $\bigcirc$ an empty cut;
6. there are three conditions in C, identical for all graphs. The first two conditions apply only to areas and cuts respectively. The third condition guarantees that characters of the same type (i.e. tokens of the same variable) are assigned the same value. The conditions restrict all models at p for all $p \in P$ in the following way:

   (a) if p is of type A then: $\begin{cases} \exists q \leq_\star p, val(q) = \bigcirc \Rightarrow val(p) = \bigcirc \\ \quad val(p) = \square \text{ otherwise} \end{cases}$

   (b) if p is of type C then: $\begin{cases} \exists q \leq_\star p, val(q) = \bigcirc \Rightarrow val(p) = \square \\ \quad val(p) = \bigcirc \text{ otherwise} \end{cases}$

   (c) $q \leq p$ and $r \leq p$ and $t(q) = H_l, t(r) = H_m$ and $l = m \Rightarrow val(q) = val(r)$.

When reasoning about models over a given graph conceived as an SI, it is clear that within any particular model the value at any part p transmits to the parts above it in the order (all $r > p$) all the logically relevant information from all the parts below it (all $q < p$) in the sense that the value at p *just is* a set of values for all $q \leq p$ that is inherited by all the values of parts above p in the model. In other words, a model m with value x at p for any given graph G will be equivalent to a model m' with value x at p' of any graph G' that differs from G only below p.

This means that a model at p^* for any graph amounts to a complete determination of the values of *all* the parts of the SI. Every such model corresponds to a semantic valuation of the graph that respects its logical structure under the standard interpretation of areas as conjunction, cuts as negation and characters as propositions.

This follows as a direct consequence of the way the compatibility conditions have been formulated. Any part is either a character, an area, or a cut. If it is a character or a cut, then it is necessarily directly below the area on which it is scribed and transmits its value directly to that area (along with all the other characters and cuts immediately on the same area). If it is an area, then either it is the outermost area, in which case its value represents the value of the model as a whole, or it is enclosed in a cut and thus transmits its value to the cut which in turn transmits the inverted value to the area just above it.

A simple EG_α graph serves as a useful example. The following graph represents the conjunction of premises for a *modus ponens* argument: $\{A \Rightarrow B, A\}$:

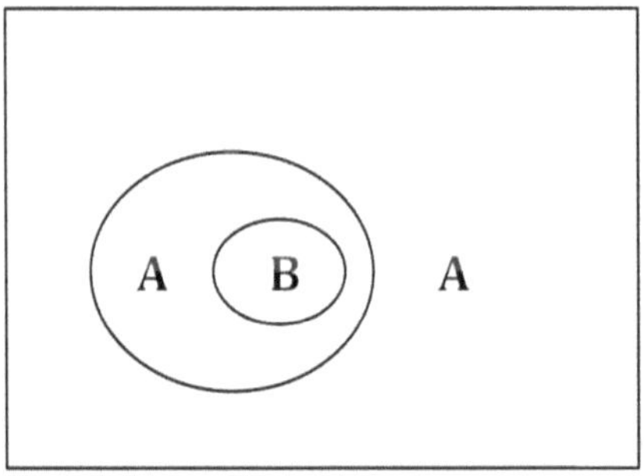

If we treat this graph as a Sheet of Indication, we stipulate first by convention that we are able to indicate any of its areas, cuts or characters. It is easy enough to do this in natural language, but one could also do so by a system of pointing, by affixing colored labels, by coding the parts in some computer language, or some other means. From a Peircean semiotic perspective, it is notable that this grounding feature of the SI conjoins a *symbolic* with an *indexical* aspect: the decision on parts and their order establishes a conventional, rule-based practice that functions to directly designate features of the diagram itself. Inspection of the graph reveals that there are exactly three distinct areas distinguished by two nested cuts, and a token of the character-type A is scribed on each of the outermost and middle areas while a token of the character-type B is scribed on the innermost area. The parts of the graph are thus ordered as in the following exploded diagram, where arrows represent the relation $\leq$ and the reflexivity and transitivity of the relation are implicit:

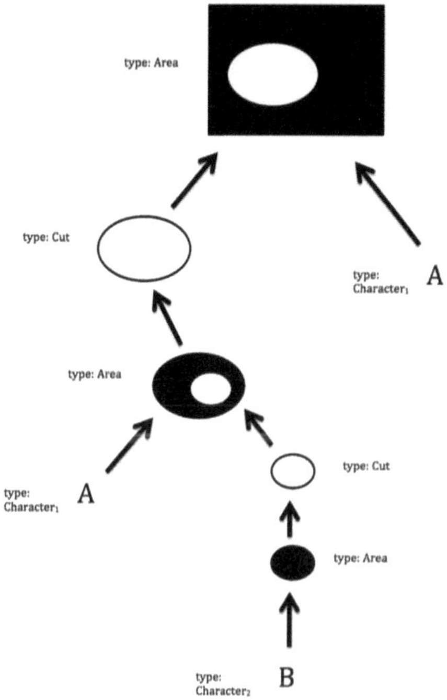

Each of these parts p may be assigned either of the two values ◯ or □ subject to the conditions in C that ensure the logical coherence of the models under the standard interpretation (graphs on the same area correspond to their conjunction; a cut negates the conjunction it contains). Since the two characters (strictly speaking, two character-types) A and B are the only variables in the present graph, there are exactly four possible models for the graph as a whole, that is to say, models defined at the outermost area. These are readily represented by diagrams that copy the original arrangement of areas and cuts and simply replace all characters in the original diagram with their assigned values in the model. The four possible valuations then appear as follows (where the dotted square is meant to signify that the substitution of the value of the empty sheet □ for a character is "invisible" in the resultant diagram):

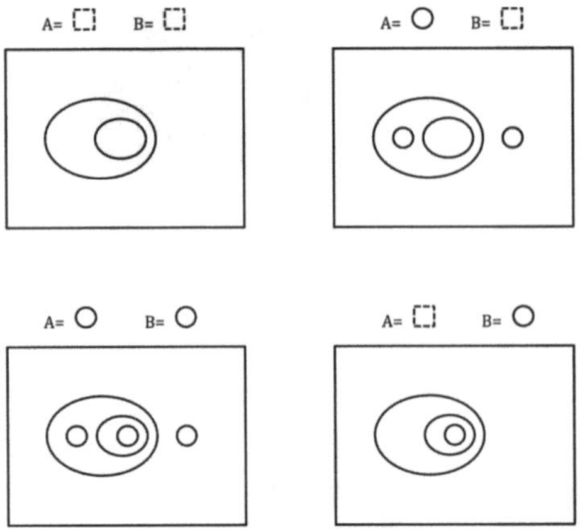

Each one of these models may itself then be conceived as a well-formed EG_α graph composed only of areas and cuts. Thus we have a function which assigns to each model at the outermost area of any graph a unique "areas-and-cuts-only-graph" that corresponds to that model's valuations of the graph's variables. Each of these latter graphs then reduces to the single value □ ("true") or ○ ("false") according to the interesting and elegant diagrammatic calculus developed in [18]. The deep connection between this calculus and Peirce's alpha-graphs has been elaborated in [6]. In the current semantics, the truth-value of the graph as a whole may be simply read off of the value of the outermost area in whatever model.

By no means is the formalization above meant to represent the only possible way to translate alpha-graphs into Sheets of Indication. However, it does serve as a useful and efficient base model that is reasonably simple in structure while still accommodating the necessary logical dynamics of the standard semantics and also accounting for a wide variety of diagrammatic usages of the graphs. It is nevertheless possible to enrich the modelization, for instance by including additional parts such as *collections* of areas, cuts and characters for any given graph. It is also possible to simplify the modelization in various ways in order to focus attention on specific features of the graphs for the purpose of investigating and exploring such features diagrammatically and pragmatically. Experimentally reducing or augmenting the order of selectable parts p or the set of conditions C induces changes in the semantic models μ_p. Such changes may be tracked, studied and then used as the basis for abductive inferences and experimental testing of the system EG.

For instance, we can formulate any EG_α graph as an SI in a new way by restricting the set of parts P as above to the new set $P' \subset P$ consisting of only the outermost area (the terminal part) and the graph's characters (thus "forgetting" the cuts and all nested areas as selectable parts taking values in the semantics). The relation $\leq$ defined over this smaller set of parts is allowed to remain unchanged (subtracting one or more elements from a partially-ordered set does not alter the partial order for the remaining elements), and the types and type-assignments need not be modified.

We then enlarge the domain of interpretation D to D' where the elements of D' are all well-formed EG_α graphs, and we stipulate the pair of conditions C':

- p is of type $A \Rightarrow val(p) = \Box$.
- $q \leq p$ and $r \leq p$ and $t(q) = H_l, t(r) = H_m$ and $l = m \Rightarrow val(q) = val(r)$.

This amounts to the reduction of the set C above to merely its last condition. The assignment of the empty sheet to the outermost area serves here as no more than a kind of dummy variable. The outermost area of the graph is necessary as an element of P' merely to ensure that the SI has a definite terminal part.

An SI semantics is then fully defined that treats any given EG_α graph G as a diagram-schema in which the characters scribed on the SA of G are replaced by entire graphs (with character-tokens of a single type replaced by the same graph). In the example above, for instance, we can replace the characters A and B with individual graphs, as shown in the picture below.

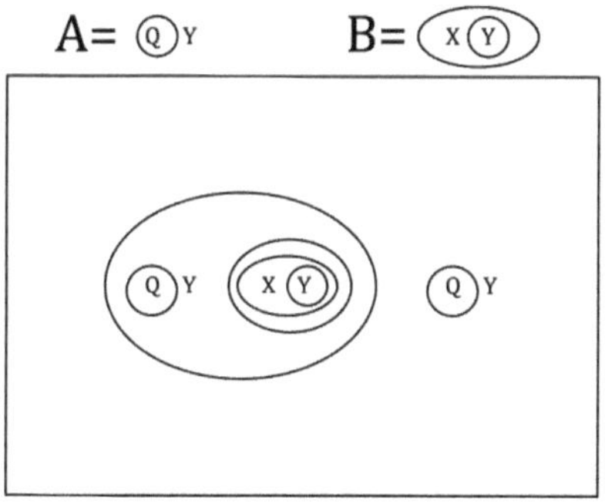

Thus each model at the outermost area according to the revised semantics corresponds to a well-formed alpha-graph that may be taken as an instance of one pair of premises for a formal *modus ponens* argument. In fact, graphs of *all* such instances are generated by the models in the semantics. Any one of these may be selected and then itself interpreted as an SI according to the previous semantics. This will then generate in turn all the models for that graph corresponding to possible truth-values assigned to its characters.

In this way, the semantics of the Sheet of Indication provides a kind of multi-use tool for investigating one of the most interesting features of Peirce's graphs, namely the fact that a single graph can be understood both as a particular logical formula and also as a general schema for classifying all graphs sharing a certain diagrammatic "shape". Hands-on users of Peirce's EG are undoubtedly aware of the utility of being able to recognize such familiar "shapes" in complex graphs. Indeed much of the practical appeal and applicational promise of EG derives from this iconic feature, which itself warrants more extensive investigation (drawing especially on the work of [16]). Yet any analysis of the graphs' iconicity – however exact and correct in its own way – that relies only on sentences formed in natural and standard formal languages risks explaining the phenomenon at issue only by occluding it in its own explanatory workings. The diagrammatic semantics of the SI provides a possible route for examining and extending the key iconic features of the graphs in an equally iconic way: here, the *explanans* is itself in continuity with the *explanandum* without thereby inducing a vicious circularity. Such an approach would seem to accord with Peirce's own rather special philosophical orientation in which issues of logic and mathematics, phenomenology ("phaneroscopy"), metaphysics, semiotics, aesthetics and ethics are intrinsically integrated via the topological notion of the continuum [19].

Peirce developed the system of EG in order to clarify and extend the powers of reasoning deductively from arbitrary hypotheses, that is, "mathematics" in Peirce's broadly conceived sense. For Peirce, mathematics investigates the domain of relations as such, and the proper tool for such investigation is the diagram. He writes ([20] 4.530 cited in [21], p. 191):

> [The object of mathematical reasoning] is *the form of a relation*. Now this form of relation is the very form of the relation between the two corresponding parts of the diagram. For example, let f_1 and f_2 be the two distances of the two foci of a lens from the lens. Then
>
> $1/f_1 + 1/f_2 = 1/f_0$
>
> This equation is a diagram of the form of the relation between the two focal distances and the principal focal distance; and the conventions of algebra (and all diagrams, nay all pictures, depend upon conventions) in conjunction with the writing of the equation, establish a relation between the very *letters* f_1, f_2, f_0 regardless of their significance, the form of which relation is the *Very Same* as the form of relation between the three focal distances that these letters denote.

In light of such statements it should be clear that the Peircean philosophical program requires that attention be paid to how the "form of relation" between parts of diagrams may be identified with the "form of relation" between elements in various domains, and this in turn would suggest the need for developing good analytical tools for tracking and extending our capacities to use diagrams in the course of inquiry. The Sheet of Indication provides an instrument both for clarifying how diagrams function in multiple contexts and for experimenting with diagrams themselves. In particular, the Sheet of Indication charts new possibilities for probing the still largely unexplored territory of the Existential Graphs.

3.5 Logical Consequence in Peirce's Transformation Rules

The main advantage of conceiving the semantics for the Sheet of Indication in terms of pre-sheaves over partial orders is that it provides a natural framework for thinking of semantic part-whole relations, and thus compositionality, over the different parts of a diagram. The canonical mathematical model of pre-sheaves over a partial order is that of continuous functions over the open sets of a topological space. Interpreting the EG_α graphs from the standpoint of the Sheet of Indication thus aligns the semantics for Peirce's notation with the mathematics of presheaves.

The reader has perhaps already noticed that the Sheet of Indication is defined in such a way that every part of a SI (that is, every p element of P) taken together with all the parts $q < p$ is also itself a SI with the domain of interpretation, types and type-assignments and compatibility conditions held invariant (NB: technically, the type-assignments function must be restricted to the elements $q < p$). In this way, every SI may be conceived naturally as a macro-level "sheet of sheets", that is, a layered embedding of "local" Sheets of Indication into larger and more extensive ones that tracks the partial order defined "globally" by the ultimate layer or sheet.

Independently of the pre-sheaf structure specific to the Sheet of Indication, it is clear from straightforward model-theoretic semantic considerations that the following relation holds between any two graphs G and G':

Given a function f that assigns to each true model of G the set of all the models of G′ that do not differ in value at any character-type (i.e. variable) that is scribed somewhere on both G and G′, then if and only if all the models in all the sets in the image of f are true, G′ is a logical consequence of G.

Informally, if the values of all the variables in a logical graph G are fixed by some true model m, and if we then transfer those fixed values to any variables found also in some other graph G' while allowing all other variables in G' to vary freely, and in so doing we generate a model m' of G' that is false, then G' cannot be a logical consequence of G since there would then exist at least one possible assignment of values to the variables in G and G' which makes G true and G' false, that is, a model in which G is true and G' is false. Yet if we range over all true models of G in this way and never find such a counter-instance, then no model exists in which G is true and G' is false; therefore G' is a logical consequence of G.

While it is perfectly general in scope, this result is unfortunately not often helpful in practical contexts, such as actually trying to determine if a given graph is a logical consequence of another graph. If one is already in a position both to know all the possible models of the two graphs independently and to be capable of constructing the relevant function f, one usually has much more than the minimal information sufficient to determine the matter. Any inquirer already in possession of such an abundance of information is not likely in fact to be in a position of what Peirce calls "genuine doubt" with respect to the question of logical consequence in any particular case.

Such considerations account in large measure for the practical utility of Peirce's graphical notation and its transformation rules. These latter are useful tools for actually manipulating and testing logical relations in contexts of genuine inquiry. Their logical soundness is of course a necessary condition for their usefulness, but their practical utility is not an immediate consequence of their soundness. What makes them genuinely useful (if they are such) is the extent to which they simplify and abet actual processes of reasoning, for example by suggesting abductive hypotheses and deductive strategies to those engaged in logical reasoning.

Given a graph G with the order P of its parts, let us define the *augmentation of* G as the graph G' obtained by inscribing another graph X on an area of G (with the order P' of its parts). This induces a natural embedding P into P': to each element $p \in P$ we associate the element $p' \in P'$ which is simply the same element p embedded into the augmented graph G'. Using the construction from the previous section, for each of this two graphs, we consider the set M of models at p^* and the set of models M' at p'^*. We now define a function F that assigns to each element of M a subset of M':

$$F : M \longrightarrow \mathcal{P}(M')$$

Given a model at p^*, say μ, we define $F(\mu)$ as the set of all models ν at p'^* such that $\forall p \in P$ of type H_i we have that $val_\mu(p) = val_\nu(p')$.

We now first show that the images of F are all disjoint and secondly that all elements in M' belong to some $F(p)$. Together these conditions show that F_μ induces a partitioning of M' into equivalence classes. Therefore, the map $g : M' \longrightarrow M$ that takes $\nu \in M'$ into the correspondent μ is well defined. It then becomes possible to reframe Peirce's transformation rules as model-theoretic claims about augmentation (and, dually, co-augmentation) of graphs.

Rather than working within model-theoretical structures in this way, however, a more diagrammatic approach allows us to see the transformation rules as themselves basic structures over which models of their own kind may be defined.

Consider the following pair of diagrams, which represent the two transformation rules Write Odd and Erase Even:

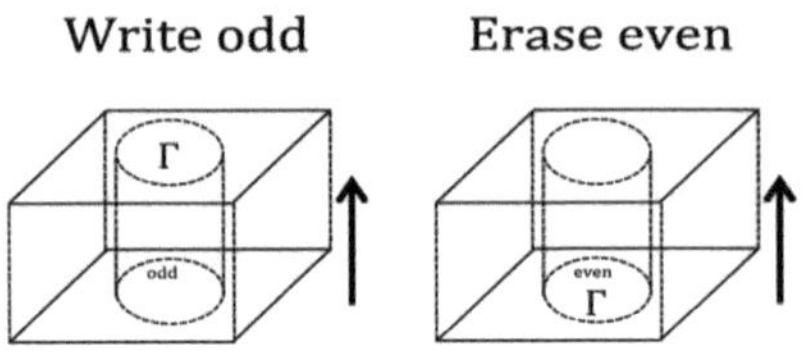

How may we interpret these diagrams as Sheets of Indication? As with the Venn diagram and playoff bracket examples from chapter two and the EG_α graphs above, it is a matter of stipulating the parts of the diagram and their types that any user of the diagram may indicate, ensuring that those parts are ordered by a partial order with a terminal part, establishing a domain of interpretation, and then establishing reasonable compatibility conditions for the models. Rather than working through all the technical details in the present case, we informally designate the relevant components of the SI as follows:

- Parts: each of the two parallelograms (lower and upper) is a part; the dotted cylinder is a part; the character Γ is a part; and the entire diagram is a part.
- The Partial Order $\leq$: character $\Gamma \leq$ dotted cylinder; dotted cylinder $\leq$ lower parallelogram $\leq$ entire diagram; dotted cylinder $\leq$ upper parallelogram $\leq$ entire diagram (with the evident reflexive relations and transitivities); no other relation holds.
- Types: {parallelogram, cylinder, Γ, whole}, with the obvious function from parts into names.
- Domain of Interpretation: $\{EG_\alpha graphs\} \cup \{components\ of\ EG_\alpha\ SIs\}$
- Conditions:

1. p is of type $parallelogram \Rightarrow val(p)$ is an EG_α graph.
2. p is of type $cylinder \Rightarrow val(p)$ is an oddly enclosed area on an EG_α graph
3. p is of type $\Gamma \Rightarrow val(p)$ is a subgraph.

Diagrams for the remaining transformation rules may be formulated along similar lines. Since Double Cut Write/Erase and De/Iterate are bidirectional rules, we may represent each of them with a single diagram, with a simple inversion of the direction of transformation. The following diagrams represent these two double rules:

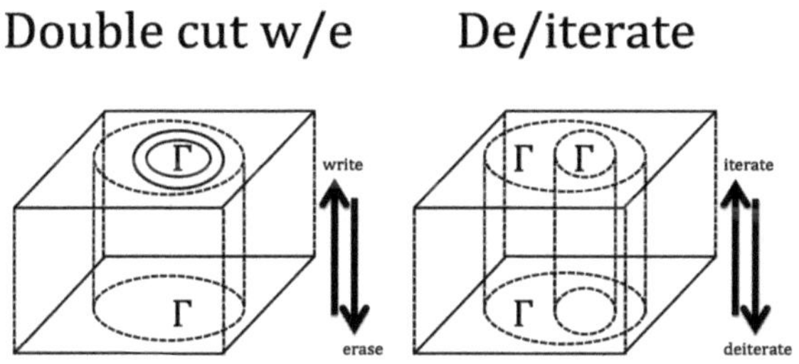

Sheets of Indication may take other sheets as elements of their models. We can build up sequences of transformation rules as in the generic figure on the left below. Here, the parts are the circles, the arrows and the diagram as a whole (with types {circle, arrow, whole}). For any circle c that is at the head or the tail of an arrow a, we have $c \leq a$ and, of course, for any part p, we have $p \leq q$ where q is the diagram as a whole. In order to generate the desired models, to each arrow we assign one of the previous diagrams of the transformation rules and to each circle we assign an EG_α graph, with the evident conditions ensuring that the graphs assigned to the circles are indeed graphs in the models over the diagrams of the transformation rules.

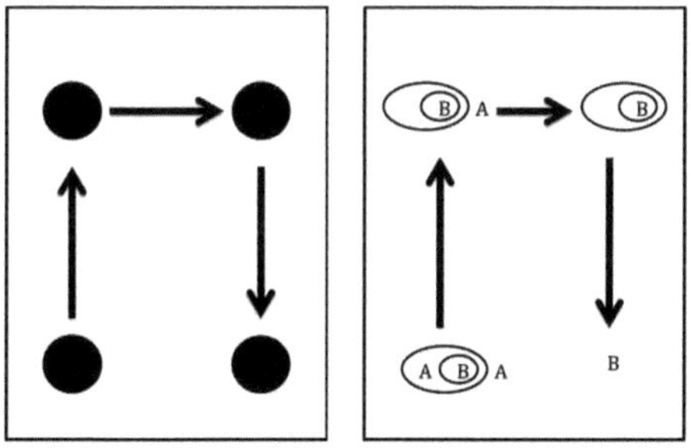

The figure on the right thus represents a particular model of the diagram on the left. By combining this model with the notion expressed above of interpreting any particular EG_α graphs as a schema for all graphs sharing its basic "shape", we may thus also treat this as a schema for any such modus ponens derivation sequence of graphs.

3.6 The Categorical Structure of EG_α

Let us denote the set of all well-formed EG_α graphs by **G**. Peirce's derivation rules depend upon a variety of relational structures among the elements of the set **G**. These relations can be formalized by considering the elements of the unstructured set **G** as the objects of a category $\mathcal{G}$. A category is a simple, highly general mathematical structure, consisting of only two components: objects and relations (arrows between objects). An elementary, yet rigorous introduction to Category Theory can be found in [22], while we refer to [23] for a more comprehensive treatment.

We define categorial objects and arrows in the category $\mathcal{G}$ as follows:

- *Objects*: well-defined graphs (elements of **G**)
- *Arrows*: given any two objects (i.e. two graphs) G_1 and G_2, an arrow exists $G_1 \rightarrow G_2$ iff G_2 may be derived from G_1 via finite application of the transformation rules given above.

It is an easy matter to check that $\mathcal{G}$ satisfies the axioms that define a category, namely

1. *Composition*: Given two arrows $f : G \rightarrow G'$, $g : G' \rightarrow G''$ there always exists a unique arrow $g \cdot f : G \rightarrow G''$ (sequences of applications of the transformation rules can be composed);
2. *Identity*: For any graph G there always exists a unique arrow $1_G : G \rightarrow G$ such that $1_G \cdot f = f = f \cdot 1_G$ for any arrow f (every graph is derivable from itself by applying no transformation rule at all);
3. *Associativity*: Given three arrows $f : G \rightarrow G', g : G' \rightarrow G'', h : G'' \rightarrow G'''$, then the composition $(h \cdot (g \cdot f))$ is the same as $((h \cdot g) \cdot f)$ (sequences of applications of the transformation rules are associative).

More details of these axioms are given in the first Appendix to this book. Because of the way we have defined our category $\mathcal{G}$, it is clear that there is at most one arrow from any graph G_1 to any other graph G_2. Even if there are multiple ways to derive G_2 from G_1 through different sequences of application of the transformation rules, this does not imply multiple arrows from G_1 to G_2. Like the binary structure of the graphs themselves (every character of a graph is either inside or outside any given area, *tertium non datur*), either an arrow exists from G_1 to G_2 or not. This structure in turn implies that $\mathcal{G}$ is a *pre-order.*

We now use various category-theoretical tools to investigate the internal structure of $\mathcal{G}$. The most important tool we employ is that of a functor. A *functor* is a "relation-preserving" map between two categories. More precisely, given two categories $\mathcal{C}$ and $\mathcal{D}$ a functor $F : \mathcal{C} \longrightarrow \mathcal{D}$ is a function which takes any object $C \in \mathcal{C}$ to the object $F(C) \in \mathcal{D}$ and takes any arrow f between two objects $C, C' \in \mathcal{C}$ into an arrow $F(f) : F(C) \rightarrow F(C')$. We also require that:

1. Composition of arrows is preserved;
2. Identity maps are mapped to identity maps ($F(1_C) = 1_{F(C)}$) for any object $C \in \mathcal{C}$.

We will be interested in both endofunctors and exofunctors of $\mathcal{G}$.[1] An important instance of the former may be given in the following way: we first define an equivalence class on G, by declaring two graphs equivalent if they can be homeomorphically transformed into each other. This is just to say that the only relevant structural features of the graphs are those concerning the combinatorial relations between enclosures and characters. For a more formal treatment see [17]. We then associate each class with a single graph that is an element of that class, using the axiom of choice. This association gives rise to a natural endofunctor $f : \mathcal{G} \to \mathcal{G}$ that "shrinks" $\mathcal{G}$ to an equivalent subcategory of itself in which every class of graphs that are "the same" are represented by a single graph. We may in a similar fashion define another class of functors that reduce $\mathcal{G}$ from a pre-order to a *partial order* (thus collapsing syntactically equivalent graphs).

Such endofunctors correspond to various scales of viewing the graphs, various Gestalts of semantics and syntax. The functorial nature of the transformation ensures that the logical relations of derivation are always preserved. The following pair of images displays the basic idea.

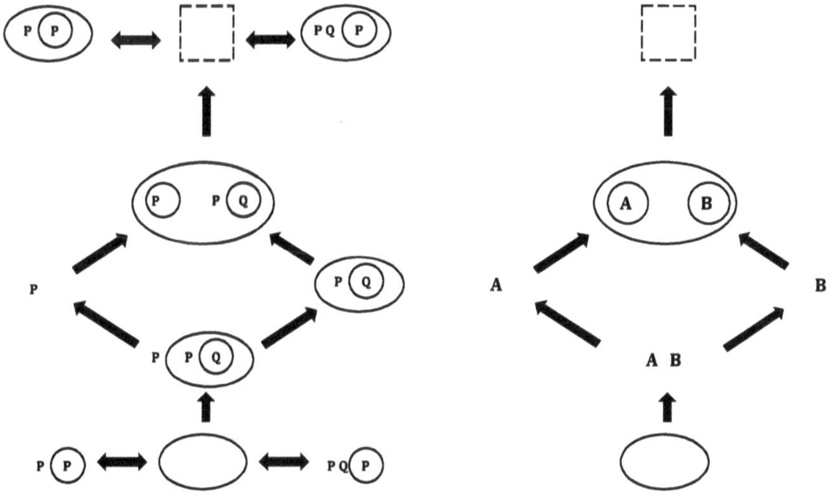

The lattice-structure of $\mathcal{G}$ exhibits various interesting properties, many of which may be defined by exofunctors into and out of $\mathcal{G}$ that correspond to determinate structural aspects of EG_α. For instance, a derivation of G_2 from G_1 corresponds to a functor $\mathbf{2} \to \mathcal{G}$, and any sequence of derivations corresponds to a functor $\mathcal{N} \to \mathcal{G}$, where $\mathcal{N}$ is the category of natural numbers with the obvious order. Most powerfully, perhaps, it provides an iconic intuition of what could be called the "flow of truth" involved in any such natural deduction system. As the diagrams above show, once what is called the *skeleton category* of $\mathcal{G}$ has been constructed, the minimal element

[1] We introduce these terms according to the following straightforward definitions: an *endofunctor* of a category C is a functor $C \to C$. An *exofunctor* of C is any functor $X \to C$ or $C \to X$ where C is not equal to X.

of the lattice (in category-theoretical terms the *initial object*) may be identified with what Peirce calls the "pseudo-graph," namely the graph with only an empty cut on the sheet of assertion which represents the value "False". By the same token, the maximal element of the lattice (the *terminal object* in the category) represents the value "True" and may be identified with the empty sheet of assertion itself. All intermediate values G such that $False \to G \to True$, where $G \neq False$, $G \neq True$, represent graphs that are true under certain conditions, namely those conditions which ensure that what they assert is in fact true. In this way, the three most general "roles" in the lattice – minimum value, intermediate value, maximum value – correspond to the modalities of impossibility, contingency and necessity respectively (as concerns the satisfiability of the graphs at issue). Indeed, to represent this in an iconic fashion, there is a natural class of functors $X : \mathcal{G} \to \mathcal{G}$ that takes the empty sheet of assertion (and all graphs isomorphic to it) to itself, takes the pseudo-graph (and all graphs isomorphic to it) to itself and all "intermediate" graphs to some single chosen graph that is neither the empty sheet nor the pseudo-graph. The "three-valued" collapses of $\mathcal{G}$ that result from these functors are all isomorphic to the category **3**, the "triangle" category that has the same structure – the same iconic "shape" – as Peirce's philosophical conception of the triadic sign.

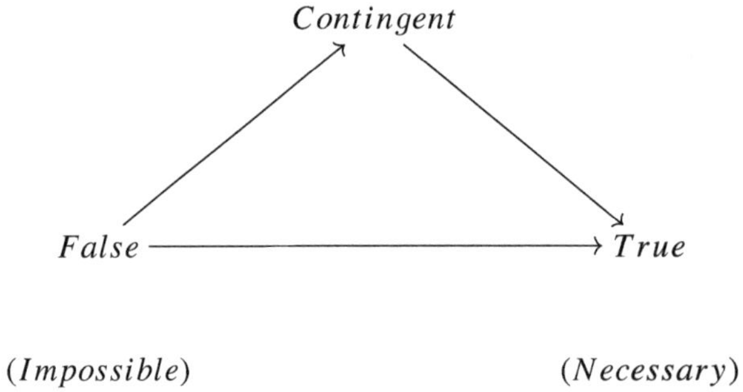

Syntax-preserving maps within $\mathcal{G}$ may be constructed so as to identify and compare visual or iconic Gestalts within the graphs. For instance, consider a representation of the sequence of transformations in EG_α that corresponds to the logical rule of *modus ponens* as compared to those corresponding to *disjunctive syllogism* (in each case, the sequence: deiterate, erase double cut, erase even):

Example – Modus Ponens:

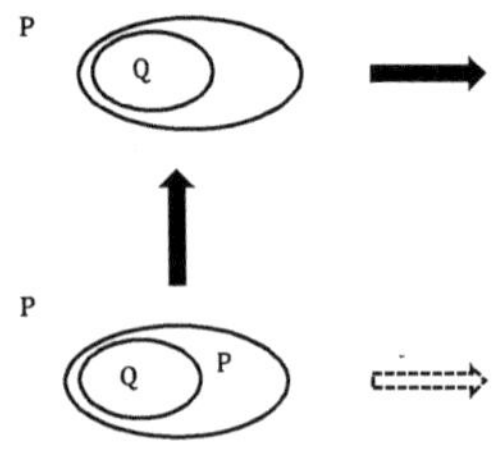

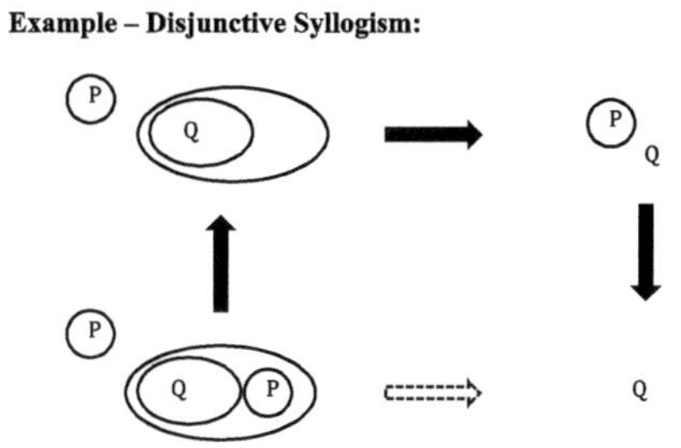

The usefulness of considering the system of EG_α within the framework of category theory consists in the fact that the iconic relations that enable one graph's "looking like" another graph (roughly, showing the same visual Gestalt) to correspond to actual common logical properties they share are representable by structures within the category $\mathcal{G}$ that may be made manifest through purely categorical constructions, such as the functors indicated above. These categorical relations themselves may be shown to possess further, sometimes surprising, iconic properties, especially when they are represented by intuitive tree-like diagrams. What this means is that the diagrammatic expression of the second-order iconicity that emerges naturally in $\mathcal{G}$ is able to process both *inherent* and *imputed* relations in the graphs, iconic semantics and algebraic syntax. Peirce summarizes these complementary kinds of diagrammatic relations as follows:

> The diagrams in which the hypotheses are embodied are of two kinds. In the one kind the parts of the diagram are seen in the visual image to have the relations supposed. In the other kind of diagrams, the parts have shapes to which conventions or "rules" are attached, by means of which the supposed relations are attributed, or imputed, to the parts of the diagrams. Geometrical figures are diagrams of the inherential kind, while algebraical formulae are diagrams of the imputations kind ("On the Logic of Quantity" p. 46–47 in [24]).

3.6.1 *Truth Values as Initial and Terminal Objects*

One of the striking features of Peirce's EG is that the Boolean "truth-values" of true and false which in more standard notation are represented by a two-element set into which syntactical strings are mapped by interpretation functions are, in EG, themselves represented by particular graphs. In other words, the graphical syntax of EG already includes, internally, the seemingly "external" binary evaluative marks of truth and falsity. The value True is represented by the empty sheet of assertion, the SA inscribed with no characters and no cuts. The value False is represented by the "empty cut", that is, the SA inscribed with no characters and exactly one cut.

The skeletal functor that collapses isomorphic objects in EG_α into equivalence classes maps all logical tautologies into the equivalence class containing the empty SA and all logical contradictions into the equivalence class containing only the empty cut. Thus all logically necessary graphs are effectively identified with the value True and all logically impossible graphs are identified with the value False. From a strictly categorical perspective these two truth-values expressed as equivalence classes of graphs are determined by their status as terminal and initial objects respectively. The proofs are trivial but instructive. Note that in what follows we will frequently make use of an abuse of (formal) notation that blurs the distinction between variables representing graphs and graphs themselves. The context in each case will obviate any confusion. Indeed, the intuitive ease of working with such "abuse" of notation should lead the reader to recognize the unique power of iconic representation in logic.

In the following diagrams, an arrow between two graphs will represent the transformation of one graph into another according to one of Perice's transformation rules defined earlier in the chapter. We will label the arrows as follows:

- EDC will denote *Erase Double Cut*
- WDC will denote *Write Double Cut*
- EE will denote *Erase Even*
- WO will denote *Write Odd*
- IT will denote *Iterate*
- DIT will denote *De-iterate*
- SUB will denote the substitution of a graph G with a graph G' on an even area, allowable assuming that $G \longrightarrow G'$ (Theorem 1).

Definition 3 In a category C an object A is an *initial object* iff for any object T of C there is exactly one arrow $A \longrightarrow T$.

Theorem 2 *The equivalence class containing the graph of the empty cut is an initial object for the category $\mathcal{G}$.*

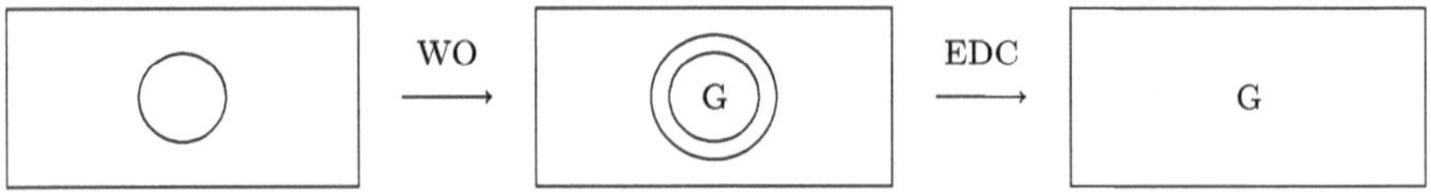

Definition 4 In a category C an object A is a *terminal object* iff for any object T of C there is exactly one arrow $T \longrightarrow A$.

Theorem 3 *The equivalence class containing the empty SA is a terminal object for the category $\mathcal{G}$.*

Proof

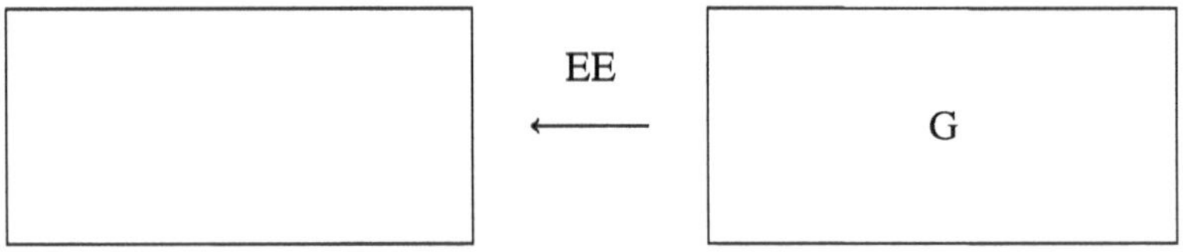

To be completely explicit, given any graph G, G may be transformed in one step into the blank SA since everything inscribed on the outermost even area (the SA itself) may be erased as licensed by transformation rule *Erase Even*, leaving only the blank

SA. Because an arrow from an object A to an object B exists in the category $\mathcal{G}$ iff A may be transformed into B in some finite number of applications of the transformation rules, an arrow from G to the blank SA exists. $\mathcal{G}$ is a preorder; therefore for any two objects C and D, at most one arrow exists from C to D. Thus, the blank SA is a terminal object in the category $\mathcal{G}$.

3.6.2 Logical Conjunction as Categorical Product

More interestingly, the logical operation of conjunction is expressed in Peirce's EG_α as the gluing together of the graphs corresponding to the logical conjuncts. That is, given two graphs G and G', the conjunction G and G' is represented by the inscription of G and G' on a single SA. This fact is true both globally (on the blank SA) and locally (on any graph area). In this sense for Peirce's EG_α mere juxtaposition of graphs in a common topological neighborhood is equivalent to their logical conjunction.

Conjunction in Peirce's EG_α corresponds to the categorical notion of product, which is given, as usual, in a limit-like fashion:

Definition 5 In any category C the *product* of two objects A and A', denoted by $A \times A'$ is an object endowed with two arrows $p_1 : A \times A' \to A$ and $p_2 : A \times A' \to A'$ such that, for all objects T with two arrows f and g respectively into A and A', there exists a unique morphism $h : T \to A \times A'$ that makes the diagram below commute ('!' signifies uniqueness).

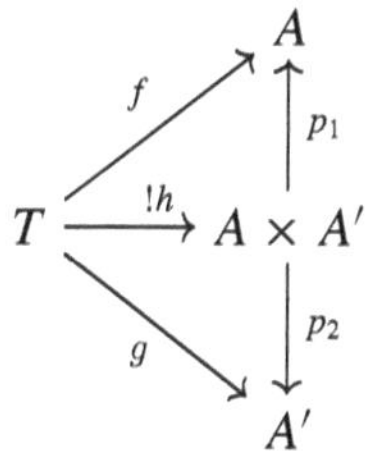

It is instructive to see what categorical product corresponds to when instantiated to $\mathcal{G}$.

Theorem 4 *Given two objects in $\mathcal{G}$, say A and B, the product $A \times B$ corresponds to the juxtaposition or "gluing" of A and B, namely the graph AB.*

Proof We need to show that for any graph X below A and B, it is the case that an arrow exists from X to the juxtaposition of A and B. It is clear that A and B may each separately be derived from AB using *Erase Even*. Since by hypothesis there is an arrow from X to A and an arrow from X to B, this means that derivations according to Peirce's transformation rules of A from X and of B from X exist. What then needs to be shown is that given two such derivations, it is necessarily the case that a derivation of AB from X also exists. The diagram below illustrates this, where the existence of the dashed arrow is what must essentially be proven (its uniqueness, if it exists, is guaranteed by construction of the category $\mathcal{G}$).

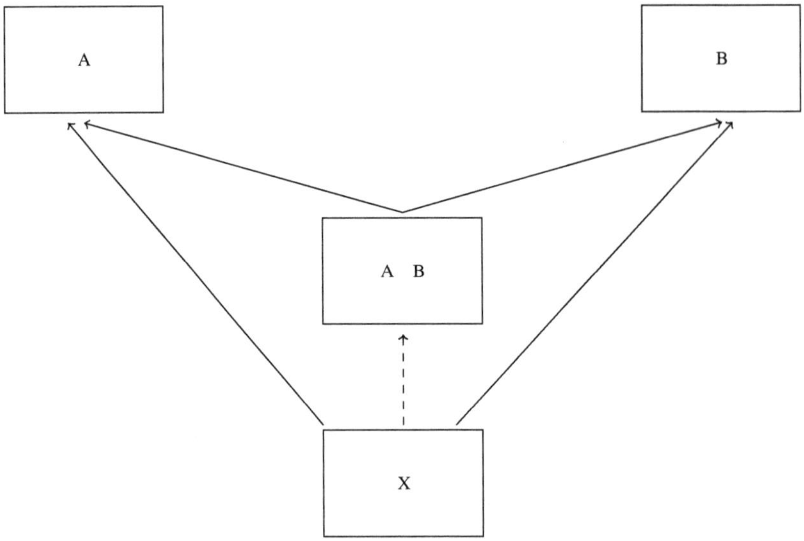

The diagrammatic proof of this is given below:

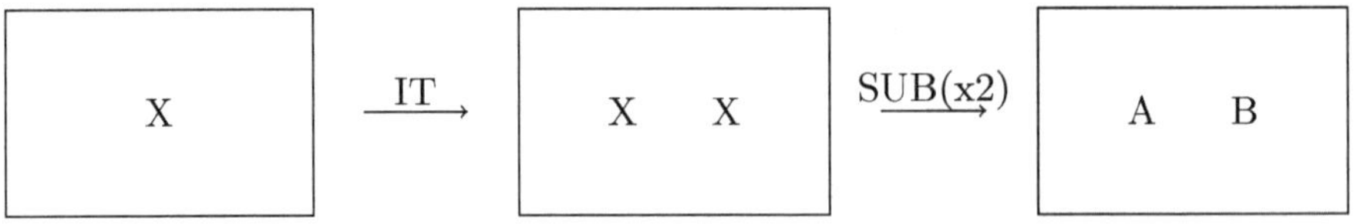

3.6.3 *Logical Disjunction as Categorical Coproduct*

Most categorical constructions have a dual form that is generated by inverting all the arrows in the definition. Conventionally, these dual constructions are assigned the prefix "co-" to distinguish them.

For instance, dual to the definition of product is the definition of coproduct as follows:

Definition 6 In any category C the *co-product* of two objects A and A', denoted by $A \sqcup A'$ is an object endowed with two arrows $p_1 : A \sqcup A' \leftarrow A$ and $p_2 : A \sqcup A' \leftarrow A'$ such that, for all objects T with two arrows f and g respectively from A and A' into T, there exists a unique morphism $h : T \leftarrow A \sqcup A'$ that makes the diagram below commute.

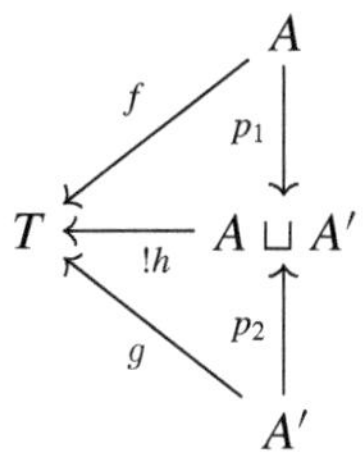

As before, we can ask the question of what is the correspondent object in the category $\mathcal{G}$. In this case, the coproduct of two graphs expresses their logical (inclusive) disjunction. We have the following:

Theorem 5 *The co-product of two graphs A and B is the following graph:*

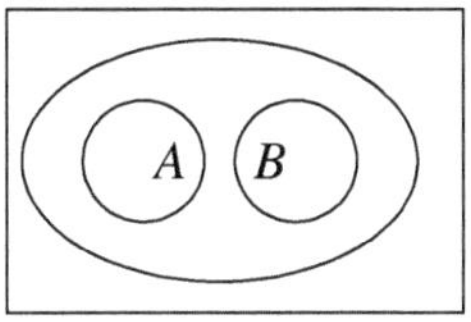

Proof Dually to what we did for the product, we need to show that for any graph X above A and B (that is, such that there exist arrows $A \longrightarrow X$ and $B \longrightarrow X$), it is also the case that X is above the coproduct $A \sqcup B$ (there is an arrow $A \sqcup B \longrightarrow X$), as illustrated below:

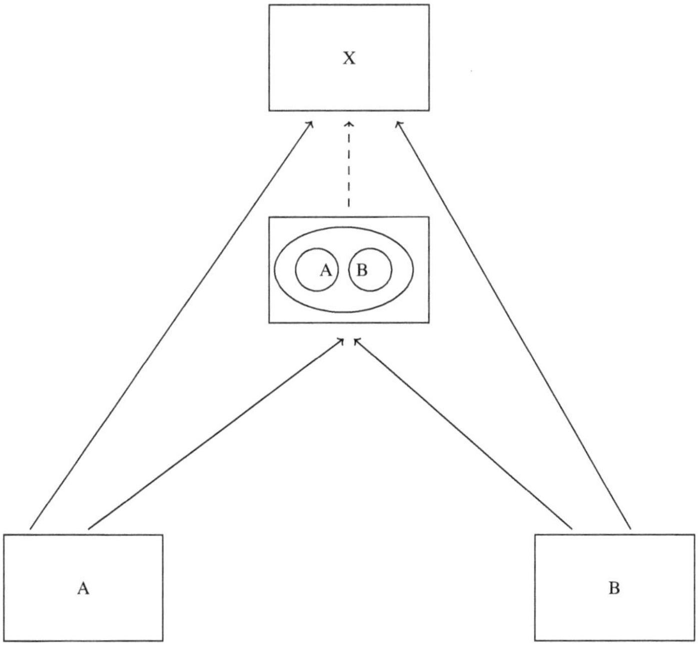

A diagrammatic proof is presented below:

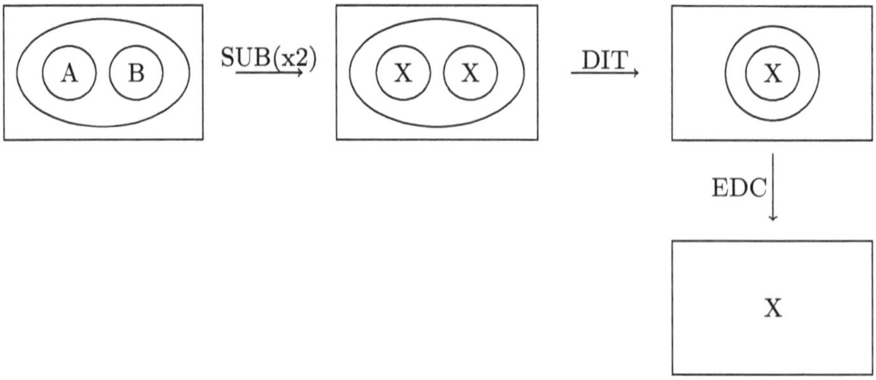

3.6.4 Logical Implication as Categorical Exponentiation

Finally, the logical operation of material implication is expressed in Peirce's EG_α as the "scroll" graph, pictured below. Just as with the truth-values True and False and the logical operations of conjunction and disjunction, the structural role of material implication may also be determined purely categorically within the category $\mathcal{G}$. In the case of implication, the relevant categorical construction is *exponentiation*. The formal categorical definition of exponential objects may be found in the second appendix (on topos theory). Below, we present the concrete instantiation of this construction within $\mathcal{G}$. The example is at once a particular case of the more general categorical structure and a helpful intuition of its essential shape.

Theorem 6 *Given two graphs A and B, the exponential object B^A is given by the following graph:*

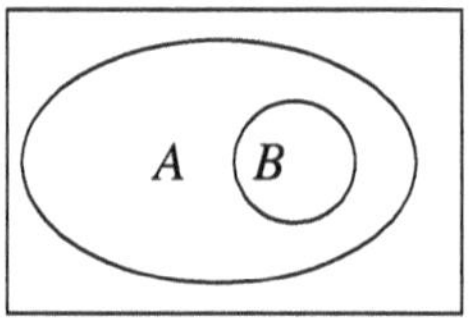

Proof We need to show that for any graph X such that the conjunction (juxtaposition) $X\ A$ is below B

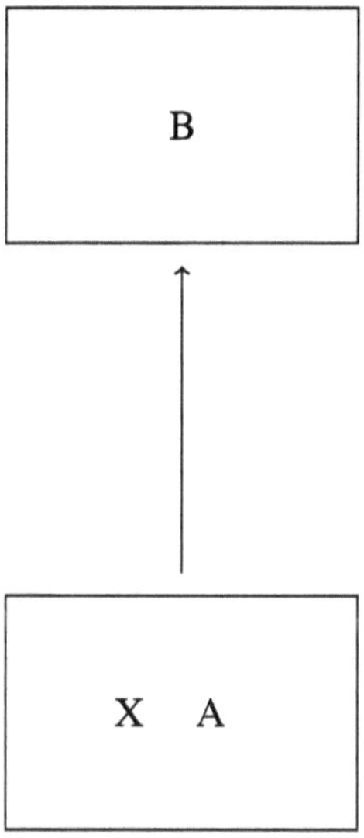

it is also the case that X is below the exponential object:

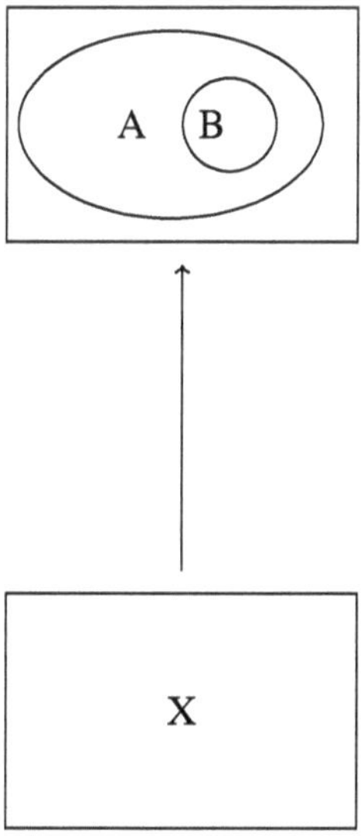

A diagrammatic proof of this is given below:

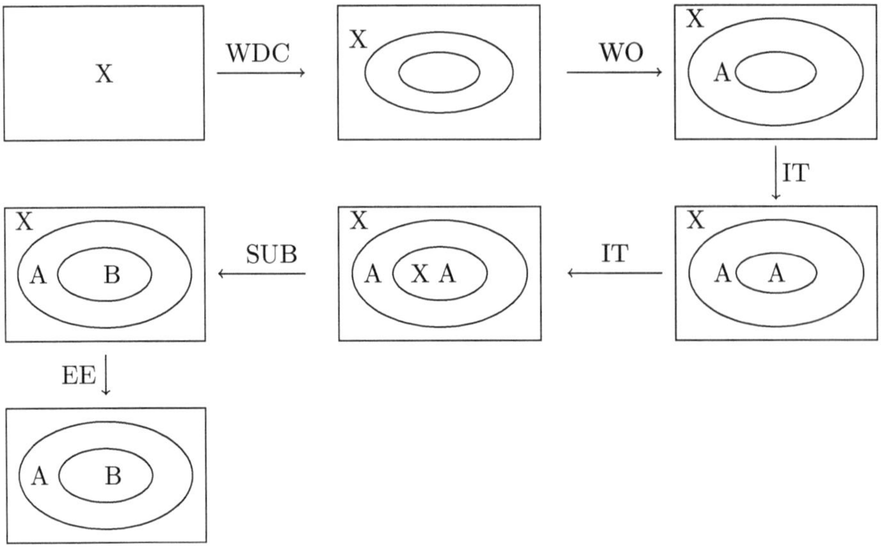

3.7 Abductive Reasoning in EG_α Via Categorical Adjunctions

Peirce's system of EG_α has been proven sound and complete with respect to classical propositional logic (CPL). It is thus formally equivalent to the various more familiar conventional notations and rule-systems for CPL, and there seems to be little reason to seek new, strictly logical results from an equivalent system regardless of its specifically iconic properties. Nonetheless, when considering in particular the creative, non-monotonic and generally problematic aspects of abductive reasoning in the framework of propositional logic, such iconic features can and do take on a relevance and importance that goes beyond mere formal isomorphism of deductive structure.

Abduction, in the framework of classical propositional logic, has been investigated quite thoroughly by Aliseda [3] among others. In that relatively simple and austere context devoid of quantifiers, one straightforward way to model abduction is by taking a set of propositions Θ (the underlying theory), a "surprising" proposition ϕ that must be accounted for, and then aiming to identify an optimal proposition (or set of propositions) α, such that when α is adjoined to Θ then ϕ can be deduced from $\Theta \wedge \alpha$. It is also reasonable to request that ϕ cannot be deduced directly from either Θ or α alone. This simple but highly suggestive model admits of many interesting variations and restrictions (see [3] for details). In the present section we show how a schematic representation of this "standard model" of abduction fits in the category theoretical framework and discuss how the categorical notion of adjoint functor is able to capture a fundamental duality intrinsic to abductive inference in general (for a more general treatment of abduction and categorical adjunction, see [25]). This duality is nicely captured by "translations" of iconicity to and from the external, categorical level and the internal, topological structure of the EG_α graphs themselves.

Let us begin by giving a diagram which illustrates the standard model introduced above (here the *absence of an arrow* between two objects is indicated by a *dotted* arrow):

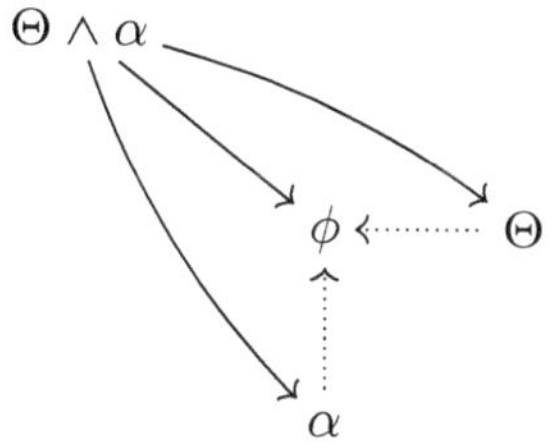

Formulated in this way, abduction may be recast in terms of a determinate diagrammatic shape within our category $\mathcal{G}$, a shape in which Θ and ϕ are given and candidates for α may vary across all possible objects which "fit" into the shape in the way displayed. The optimality property mentioned above (which makes α the *best explanation* for ϕ) then corresponds to one of the most fundamental and useful concepts in category theory, namely the *universal mapping property* (UMP), several examples of which we have already examined above. A clear illustration of a UMP can be given by considering once again but from a slightly different point of view the *product* of two objects in a given category:

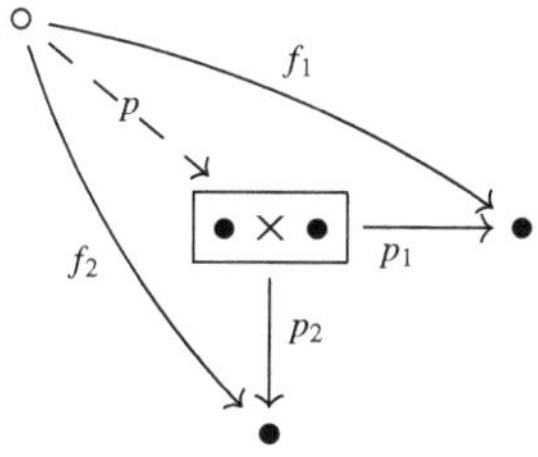

This diagram iconically represents the following: given a fixed shape of arrows (in this case a "fork" of two arrows coming out of any single object – represented here by the open circle – and going into two fixed objects) we look for the optimal candidate, that is, the special object that is capable of representing all such objects, which it does by having the property of all such candidates relating to it in a unique way. In other words, there is exactly one arrow from each of the possible candidates into this optimal one. We represent the unicity of such a relation here with a dashed arrow.[2]

The existence of objects with universal mapping properties may, in certain cases, be represented solely through functors. A functor can be seen as a generalization of the concept of function: functors not only take objects to objects, but their action also preserves relations between objects in the translation from the "domain" category to the "category". This movement opens up a rich universe of relation that cannot be seen at level of functions.

One of the most fundamental of such relations is called "adjointness", and it concerns a collection of carefully determined relations between two functors. The formal definition of adjointness can be found in any standard Category Theory textbook (see, for instance, [23]). Although a thorough discussion of adjointness goes well beyond our present scope, we believe that it is important to outline the general features of such a construction, with the goal of showing how well it fits the relational framework sought in Peirce's works for conceiving abduction.

Given two categories $\mathcal{C}_1$ and $\mathcal{C}_2$ and two functors $F : \mathcal{C}_1 \longrightarrow \mathcal{C}_2$ and $G : \mathcal{C}_2 \longrightarrow \mathcal{C}_1$, we can illustrate adjointness suggestively by the following diagram:

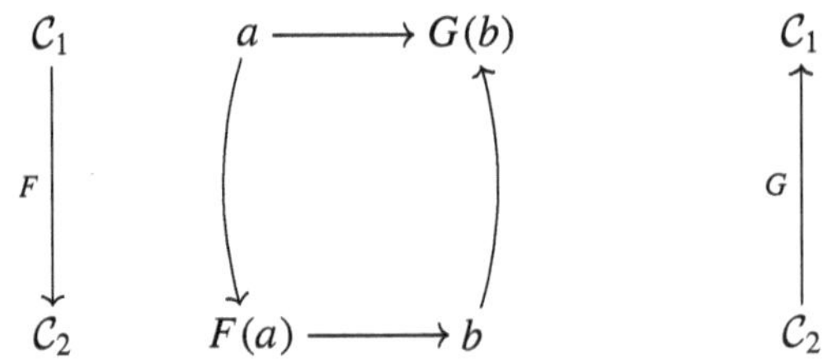

As the above picture suggests, the main idea behind adjointness is to constrain the images $G(b) \in \mathcal{C}_1$ of the elements $b \in \mathcal{C}_2$ that are related (by an arrow!) to the image $F(a)$ of an element $a \in \mathcal{C}_1$ to be related to the "original" element $a \in \mathcal{C}_1$! This beautiful movement has an intrinsic orientation, which justifies the fact that we say that F is the *left adjoint* of G.[3]

[2]Unicity is needed when dealing with categories more general than pre-orders, in which multiple arrows between object are allowed to exist.

[3]There is a second condition, called *naturality*, that F and G need to meet in order to be adjoint. We omit discussion of it here and refer the reader to [23, 26].

The diagram below aims to further reinforce this concept:

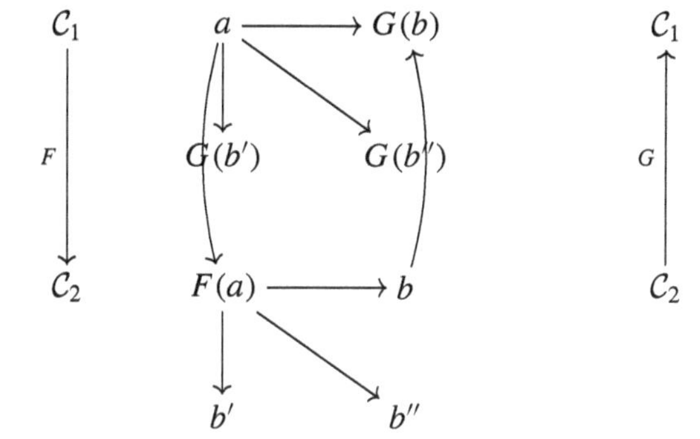

It is the case that for any category $\mathcal{C}$ for which the categorical product of any two objects in $\mathcal{C}$ also exists in $\mathcal{C}$, this very fact is itself instantiated in a higher-order relation between two functors $\Delta : \mathcal{C} \to \mathcal{C} \times \mathcal{C}$ (here Δ is the *diagonal functor* $\Delta(x) = (x, x)$). In other words, the existence of a specific type of adjunction based on a given category guarantees that the inner structure of this category coheres with respect to a certain UMP as expressed universally (present throughout the category).

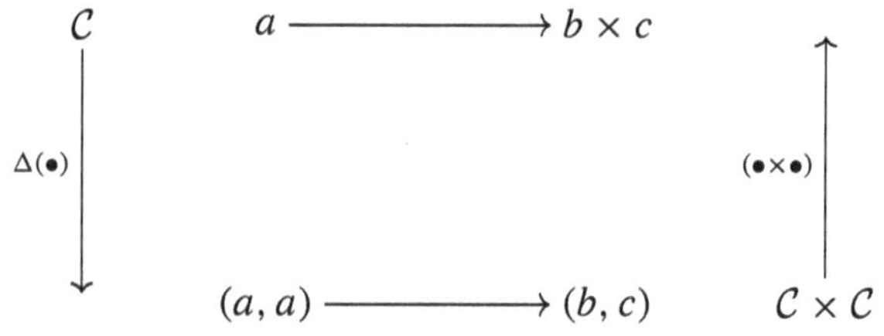

The role of adjoint functors in abductive inference has been investigated by Arzarello and others in [27]. They present two different way to attack abduction, which can be summarized by the following:

- $\Theta \vdash ? \longrightarrow \phi$
- $\Theta \wedge ? \longrightarrow \phi$

As they point out, "In both cases abduction plays an essential role in reversing the course of thought (from ascending to descending control); in the second case the use of counterexamples seems to be at the origin of a more involved course of thought, which we have called the logic of not" [27].

The duality that they identify, which is the duality between the propositional connectives $\wedge$ and $\Rightarrow$ corresponds to the fact that $\wedge$ is the left adjoint of $\Rightarrow$, as the following diagram shows:

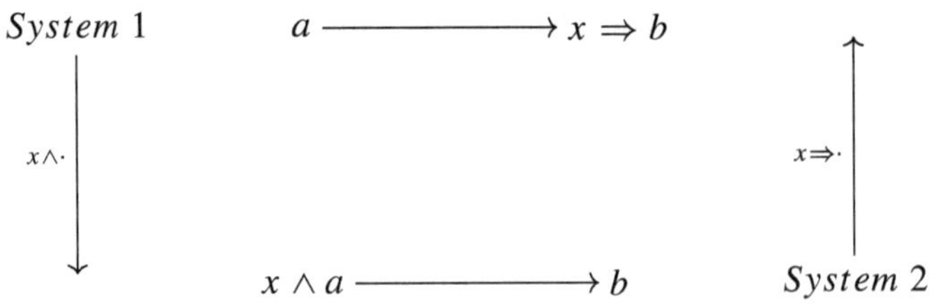

What is more striking, however, is how the above relations translate into an adjointess in EG_{α}, as shown below (in the following diagrams, rectangles – otherwise noted – represent cuts drawn on the empty Sheet of Assertion):

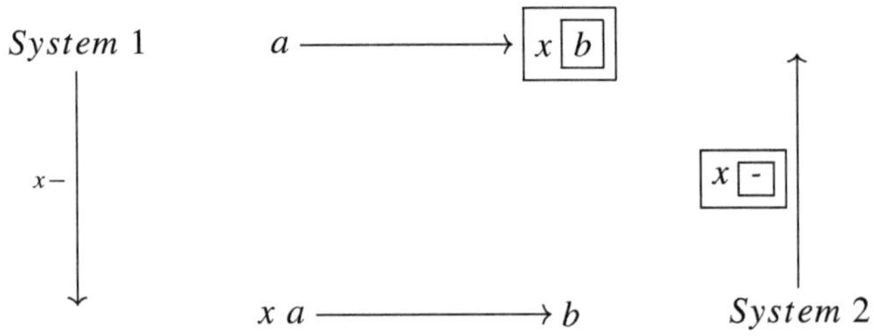

In order to prove that the above is an adjoint square, we first suppose that

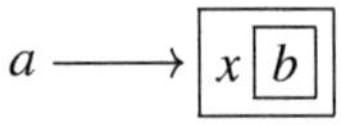

Then, using the result from the previous section along with Peirce's rules, we have that:

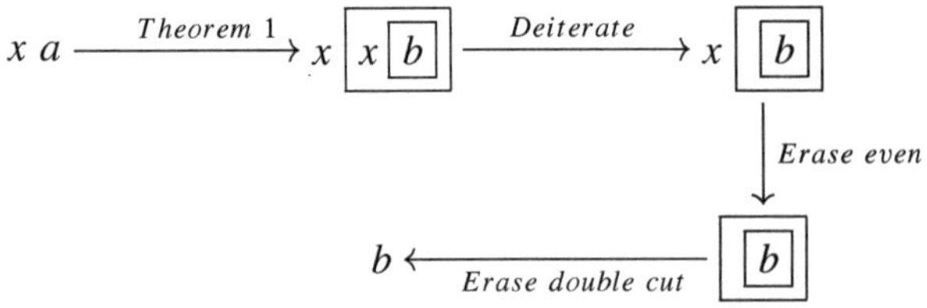

Vice versa, suppose that

$$x\ a \longrightarrow b$$

This also means that

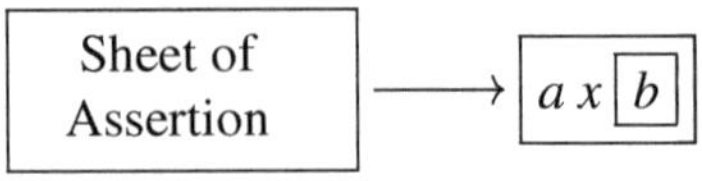

where 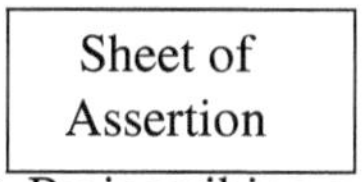

denoted the (empty) sheet of assertion corresponding to "True". By inscribing a double cut in the graph as shown below, we get

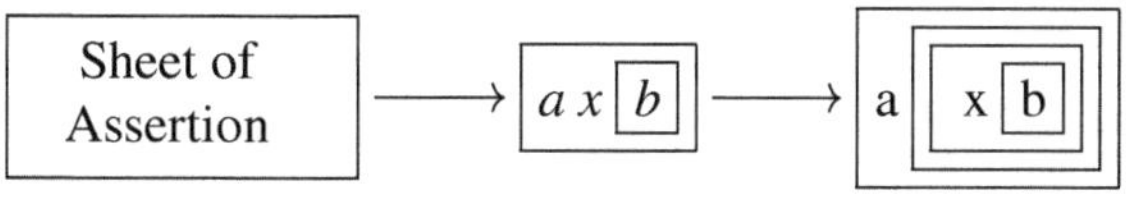

which corresponds to the statement

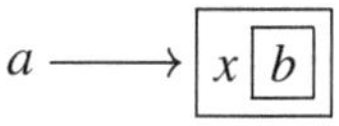

The richness of these iconic functorial relations – which, with respect to the iconic relations of the graphs themselves function as meta-iconic representations – opens a wide-ranging field for the visual presentation and intuitive comprehension of many complex structures within propositional logic.

3.8 Conclusion

Although the iconic nature of these categorical relationships allows them to be represented in a more or less immediately intuitive way, the formal machinery that is needed to work out such notions in rigorous detail can become quite intricate and forbidding for all but the most intrepid logicians and mathematicians. Some of the most fundamental elements of this work has already been done, particularly through the groundbreaking efforts of Zalamea [28] and Brady and Trimble [17], but much remains still to be constructed, studied and mathematically grounded in this regard. Part of the power of the iconicity here is that highly complex and abstract logical notions may be investigated and various hypotheses corresponding to these constructed at an initially intuitive level that is often accessible even to general academic audiences. The rigorous formal framework of category theory may then be utilized to test and ultimately ground these motivating intuitions. This entire process may itself be "captured," that is, at once recorded and represented – at least potentially – through sequences of graphs that represent various subcategories and functors iconically.

We turn now to a formal model of abductive reasoning in an alternate mathematical setting: Alain Badiou's coordination of epistemic risk and advance with the set-theoretical method of *forcing*.

References

1. W. Lawvere, Quantifiers and sheaves, in *Proceedings of the International Congress on Mathematics (Nice 1970)* (Gauthier-Villars, 1970), pp. 329–334
2. C.S. Peirce, *Selected Writings* (Dover, New York, 1958)
3. A. Aliseda, *Abductive Reasoning: Logical Investigations Into Discovery and Explanation* (Springer, Berlin, 2006)
4. D. Walton, *Abductive Reasoning* (University of Alabama, Tuscaloosa, 2005)
5. L. Magnani, *Abduction, Reason and Science - Processes of Discovery and Explanation* (Springer, Berlin, 2000)
6. L. Kauffman, Peirce's existential graphs. Cybern. Hum. Knowing **18**, 49–81 (2001)
7. A. Tarski, Sentential calculus and topology, in *Logic, Semantics, Metamathematics: Papers from 1923 to 1938. Trans J.H. Woodger* (Clarendon, Oxford, 1956), pp. 421–454
8. A. Rodin, *Axiomatic Method and Category Theory* (Springer, Berlin, 2014)
9. D.D. Roberts, *The Existential Graphs of C.S. Peirce* (Mouton, The Hague, Pairs, 1973)
10. J.F. Sowa, Peirce's tutorial on existential graphs. Semiotica **186**(1–4), 345–394 (2011)
11. F. Zalamea, Peirce's logic of continuity: existential graphs and non-cantorian continuum. Rev. Mod. Log. **9**(29), 115–162 (2003)
12. J. Zeman, Peirce's logical graphs. Semiotica **12**, 239–256 (1974)
13. A. Pietarinen, Moving pictures of thought II: graphs, games, and pragmaticism's proof. Semiotica **186**, 315–331 (2011)
14. A. Pietarinen, *Signs of Logic: Peircean Themes on the Philosophy of Language, Games, and Communication* (Springer Synthese Library, Berlin, 2006)
15. F. Zalamea, *Synthetic Philosophy of Contemporary Mathematics* (Urbanomic, Falmouth, 2012)
16. S.J. Shin, *The Iconic Logic of Peirce's Graphs* (MIT Press, Cambridge, 2002)
17. G. Brady, T.H. Trimble, A categorical interpretation of C.S. Peirce's propositional logic Alpha. J. Pure Appl. Algebra **49**, 213–239 (2000)
18. G. Spencer-Brown, *Laws of Form* (Allen and Unwin, London, 1969)
19. F. Zalamea, *Peirce's Logic of Continuity: A Conceptual and Mathematical Approach* (Docent Press, Boston, 2012)
20. C.S. Peirce, *Collected Papers of Charles Sanders Peirce* (Belknap Press, Cambridge, 1931–1935)
21. C. Hookway, *Peirce* (Routledge, London, 1985)
22. W. Lawvere, *Conceptual Mathematics: A First Introduction to Categories* (Cambridge University Press, Cambridge, 2009)
23. S. Mac Lane, *Categories for the Working Mathematician* (Springer, Berlin, 1998)
24. C.S. Peirce, *Philosophy of Mathematics* ed. by M.E. Moore (Indiana University Press, Bloomington, 2010)
25. F. Tohme, G. Caterina, R. Gangle, Abduction: a categorical characterization. J. Appl. Log. **13**, 1 (2015)
26. S. Awodey, *Category Theory*, 2nd edn. (Oxford University Press, Oxford, 2010)
27. F. Arzarello et al., Abduction and conjecturing in mathematics. Philosophica **61**(1), 77–94 (1998)
28. F. Zalamea, Categorical models for intuitionistic existential graphs. *International Category Theory Conference CT2010 lecture slides*. http://ct2010.disi.unige.it/slides/ZalameaCT2010.pdf

Chapter 4
Ontology and Abduction in Badiou's *Being and Event*

4.1 Introduction

In this chapter, we look at the problem of abductive inference as refracted through a mathematical context quite different from the topological areas and cuts of Peirce's graphs as analyzed in the previous chapter. Rather than the relational nestings of areas within graphs and the categorical structures induced in a holistic way through the relations holding among the graphs conceived as a deductive system, we will examine the model of abduction that emerges from Alain Badiou's provocative conflating of philosophical ontology and Zermelo–Fraenkel set theory in *Being and Event*.

Badiou himself summarizes the overall project of *Being and Event* in the following terms ([1], xiii):

> To think the infinity of pure multiples I took tools from Cantor's set theory. To think the generic character of truths I turned to Gödel and Cohen's profound thinking of what a "part" of a multiple is. And I supported this intervention of mathematical formalism with a radical thesis insofar as *being, qua being*, is nothing other than pure multiplicity, it is legitimate to say that ontology, the science of being qua being, is nothing other than mathematics itself.

The motivations behind the writing of Badiou's far-ranging work are clearly outlined in his Introduction to the book, and can be characterized as a response to the "flow of this ordinary philosophy" of the late 80s. Here Badiou refers to the predominant intellectual context of that period, where in his view mostly "moral philosophy disguised as political philosophy" was mainstream and "a kind of flabby reactionary philosophy insinuated itself everywhere". Badiou's own response to this perceived lack of critical rigor in philosophy is, in *Being and Event*, to use the mathematics of set theory (which he identifies—somewhat summarily—as mathematics *tout court*) as a formal domain for modeling fundamental philosophical problems concerning being, subjectivity and truth. From the standpoint of the present work, Badiou's ambitious project may be understood as developing both as and within a unique yet highly general abductive hypothesis, namely the hypothesis that *if* philosophical ontology (the

G. Caterina and R. Gangle, *Iconicity and Abduction*,
Studies in Applied Philosophy, Epistemology and Rational Ethics 29,
DOI 10.1007/978-3-319-44245-7_4

science of "being *qua* being") is identified with the mathematics of Zermelo–Fraenkel set theory *then* structures and techniques indigenous to set-theoretical mathematics will be capable of shedding light on fundamental philosophical questions in perhaps unforeseen ways.

What emerges as especially of interest when Badiou's project is seen from this point of view is that the results that he derives from *within* this general hypothesis are themselves closely aligned with the typical shape or pattern of abductive reasoning as identified earlier in chapter one. Thus, in the context of a macro-abduction about the relationship between ontology and mathematics, a potentially testable outcome results concerning the real consistency and power of abductive inferences of a particular kind.

Badiou summarizes his philosophical program in four points:

1. Situations are pure indifferent multiplicities. If truth exists it is indifferent to differences.
2. The structure of situations does not deliver any truths.
3. A subject is a militant of truth.
4. Truth is intrinsically "generic", in the sense that it "concerns everyone inasmuch as it is a multiplicity that no particular predicate can circumscribe".

With such highly politicized premises, it may be odd at first glance that Badiou turns to mathematics as the means to support his thesis. What seems to ultimately push Badiou towards the realization that "Mathematics = Ontology" is a meditation on the relation between the "discrete versus continuous" problem in mathematics—a problem which is deeply intertwined with all aspects of Set Theory, in particular with model-building.

It is precisely this latter topic that is relevant to the notion of Peircean abduction. In developing what amounts to a modal notion of truth based in non-constructible models of ZF set theory to support his claim that truth is essentially generic, Badiou is led to tackle the problem of hypothesis-formation within a dynamic view of the interplay between ontology and epistemology, with this latter being understood in Badiou's own somewhat idiosyncratic manner as the tracing of "truth-procedures". Despite its formidable technicality and extremely formal character, this way of integrating philosophical questions and mathematical structures provides insights that prove in fact to be pertinent to how abductive inference functions quite generally in the scientific method.

In what follows we will analyze the role of abduction as a truth procedure insofar as this notion serves as a fundamental piece yet also a problematic limit-concept of Badiou's formal ontology. The semiotic role of iconicity in the abductive model is here played by a particularly subtle mathematical technique: the *forcing relation* first developed by Paul Cohen to disconfirm Cantor's Continuum Hypothesis.

4.2 Abductive Reasoning in Badiou

> I term subject any local configuration of a generic procedure from which a truth is supported [1].

Badiou never mentions explicitly the term "abduction" in *Being and Event* and neglects to refer to Peirce in any significant detail in his work. At the same time, we believe that framing the entire philosophical project of *Being in Event* within an abductive perspective, helps to clarify and better understand Badiou's overarching goals and methods.

Meditation 35 ([1], pp. 391–409) deals with the relation between what he calls the "subject" and the most general notion of truth. Essentially, Badiou's thesis is that truth is not accessible from the perspective of the subject, but it can be somewhat controlled from within the subject's semantical context. Although this is a very technical point, the details of which we will develop in this chapter, Badiou offers a very informal, yet rigorous, version of such a process, drawing his example from the context of Newtonian astronomy.

Suppose—Badiou says—that a perturbation in the motion of planets leads to the formation of the hypothesis that a yet undiscovered planet is the cause of such perturbations. Badiou distinguishes three cases:

1. Given clear conceptual and logical links between the hypothesis and the observed perturbation, it is definitely true that *if* such a "perturbating" planet exists, "then the statement *a supplementary planet exists* will have been veridical in the universe constituted by the solar system supplemented by scientific astronomy."
2. Upon careful reflection, the calculations do not add up: there is no "logical path" between the term (the planet) and the hypothetical statement "something is inflecting the trajectory."
3. Despite the reasonable character of the hypothesis, no amount of observation will ever come to confirm its truth: the perturbating planet simply does not exist.

We first notice that saying that a term "exists" is equivalent to saying that such a term must eventually be "observed" or, even more suggestively "encountered".

Then case 1 establishes a definitive, modal relation between the term (the planet) and the situation-to-be (a context in which astronomy has the practical means to verify the existence of the planet).

Case 2, on the other hand, is somewhat dual to 1 in the sense that it establishes a negative truth, preventing any possible relation between the hypothetical term and the statement "something is inflecting the trajectory".

In case 3, however, a relation between the term and the statement does exist ("expert calculations allow the conclusion that this *something* must be a planet"), but nonetheless there is not, in fact, (we do not encounter) any term (planet) validating this relation.

The important point is that cases 1 and 3 do not contradict one another. It may be the case indeed that they *both* hold. This is due to the fact that case 1 takes the form of a conditional, whereas case 3 also involves a statement of (unknowable, that is, non-definitively-confirmable) fact.

The abductive nature embedded in the relations between the subject and truth is a direct consequence of the structured modalities informally described above, and we will spend most of this chapter unfolding this observation.

At the same time, this simple example shows one important (in an abductive perspective) fact: that the lack of explanatory hypothesis for a given observed fact, given a defined ground model of knowledge, does not prevent (obviously) that such an explanation can be found in an enlarged ground model.

In *Being and Event* Badiou develops a comprehensive theory of this process in a completely formal way, within the language of Set Theory. It is therefore only natural that the end point of this journey is Cohen's forcing technique, which can be regarded as the most general formalization of abductive reasoning within Set Theory.

4.3 The Continuum Hypothesis

Peirce's contemporary G. Cantor formulated what has come to be called the Continuum Hypothesis. The technique that we will be using as the *trait d'union* between abduction and the diagrammatic representation of the emergence of EG_γ from EG_β was developed by Paul Cohen in his work to prove the independence of this hypothesis from the Zermelo–Frænkel axioms of set theory. What follows is a brief summary.

Let us denote by ω the set $\omega = \{0, 1, 2, \dots\}$, by $|S|$ the cardinality of a set S, by $\aleph_0$ the cardinality of ω and by $\mathcal{P}(S)$ the power set of S. Then the Continuum Hypothesis (abbreviated CH) is the conjecture, formulated by Cantor, that, if S is an infinite subset of $\mathcal{P}(\omega)$, then either $|S| = \aleph_0$ or $|S| = \mathcal{P}(\omega)$. In particular, CH asserts that there is no set whose cardinality is strictly between that of the integers and that of the real numbers.

This conjecture was the first on the list of problems presented by David Hilbert in the year 1900. About thirty years later, Kurt Gödel [2] proved that CH cannot be disproved from the Zermelo–Frænkel set of axioms (abbreviated ZF) and that therefore, if ZF is consistent, ZF+CH is still consistent.

More precisely, Gödel defined the class of *constructible* sets and showed that the statement (which, interestingly, he denoted simply with the letter "A"):

Every set is constructible

is consistent with ZF and, moreover, that CH can be proved from ZF as restricted to constructible sets and therefore, that ZF+CH is consistent.

The intrinsic difficulties in constructing alternate models of the Zermelo–Frænkel axioms was a serious obstruction towards any substantial progress about CH. It was not until the early 1960s that Paul Cohen found a method to construct models of ZF in which neither proposition A, nor CH holds. Combined with Gödel's results, Cohen showed that CH is *independent* from ZF.

Cohen's work relies on a novel technique that he invented *ex novo*, called *forcing*. In what follows we will outline the fundamental, simple ideas behind forcing and discuss how this method constitutes a natural framework for the emergence of both of Peirce's "Logic of Continuity" as expressed in EG_γ and abduction.

4.4 The Generic Extension of Standard Models of ZF

The following section is a brief overview of the background preparation necessary to define forcing. Our presentation follows the structure that Badiou [1] uses in his own summary of Cohen.

Cohen's technique of forcing makes use of two models of ZF, one of which is taken to be the "ground model", which we will denote by M, and its *extension* $M[G]$ obtained by adjoining what will be called a *generic set* G to M. G is *generic* in the sense that it remains indiscernible from the point of view of what is codable within M, so that M and $M[G]$, while both models of ZF, function at distinct levels. Forcing will make use of this difference. The precise character of this difference is elaborated in the following sequence of steps.

4.4.1 The Generic Set

A model for the axioms of set theory can be simply thought as a "realization" of those axioms, exactly in the same way as a group, say $(\mathbb{Z}, +)$ (the group of the integers with the operation of addition), is a "realization" for the group axioms.

The goal of this subsection is to define a notion of *generic set*: intuitively, a set G is generic with respect to a model M, if "from inside" M, G can be defined but yet cannot be discerned. In order to unfold this only apparent contradictory definition, we need to outline precisely the steps of the construction. In what follows we only assume familiarity with the basic notation of set theory.

Standard Models

We start with a *standard* model M for ZF, where by *standard* we mean that:

- every element x of M is a well-founded set: that is, x is constructed inductively from the empty set using operations such as taking unions, subsets, powersets, etc.;
- M is transitive: every member of an element of M is also an element of M;
- M is countable.

Set of Conditions

Within M, we discern a set of *conditions*, that we denote by $\mathcal{C}$, such that:

- $\mathcal{C} \in M$;
- there is a partial order on $\mathcal{C}$, denoted by $\subset$.

We say that π_1 *dominates* π_2 if $\pi_2 \subset \pi_1$ and that π_1 is compatible with π_2 if there is π_3 such that $\pi_1 \subset \pi_3$ and $\pi_2 \subset \pi_3$. If we represent the order with the branches of a tree as below

then this can be expressed graphically by the following diagram:

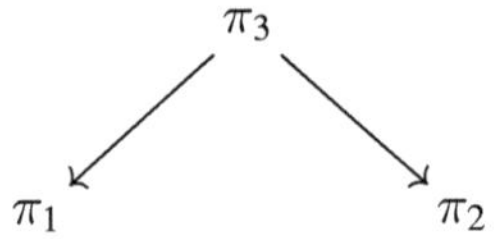

Then the last property that identifies $\mathcal{C}$ is that

- every condition is dominated by two conditions which are incompatible among themselves:

$$\forall \pi \; \exists \pi_1, \pi_2 \text{ such that } (\pi \subset \pi_1) \;\&\; (\pi \subset \pi_2) \text{ with } \pi_1 \text{ and } \pi_2 \text{ incompatible.}$$

Graphically, that means that, if π_1 and π_2 are incompatible, then a picture like the following is impossible within $\mathcal{C}$:

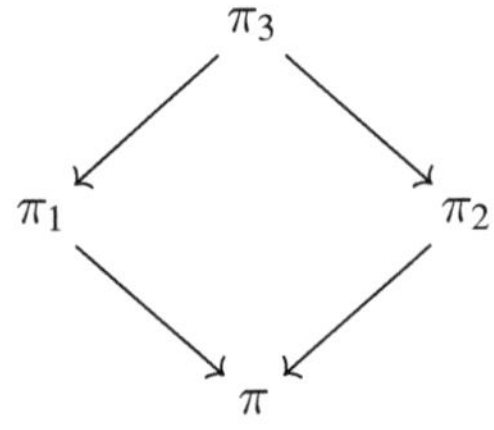

Example 3 Let C be the set of all the finite binary sequences. That is

$$\mathcal{C} = \bigcup_{n \in \mathbb{Z}^+} < a_1, a_2, \ldots, a_n >$$

where $a_i \in \{0, 1\} \; \forall i \in \{1, 2, \ldots, n\}$. Let the order be defined by

$$< a_1, a_2, \ldots, a_n > \subset < b_1, b_2, \ldots, b_k >$$

if and only if the first n entries of $< b_1, b_2, \ldots, b_k >$ coincide with $< a_1, a_2, \ldots, a_n >$; that is if and only if

1. $n < k$
2. $a_i = b_i \; \forall i \in \{1, 2, \ldots, n\}$

In this case we have, for instance, that $< 0, 1, 1 > \subset < 0, 1, 1, 1 >$. Graphically:

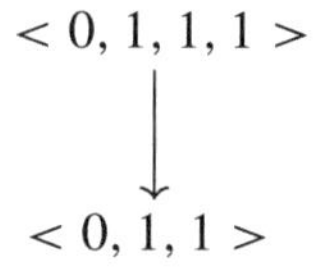

By the same token, it easy to see that, for instance, the sequences $< 0, 1, 1 >$ and $< 0, 1, 0 >$ are incompatible because there does not exist a sequence which dominates both:

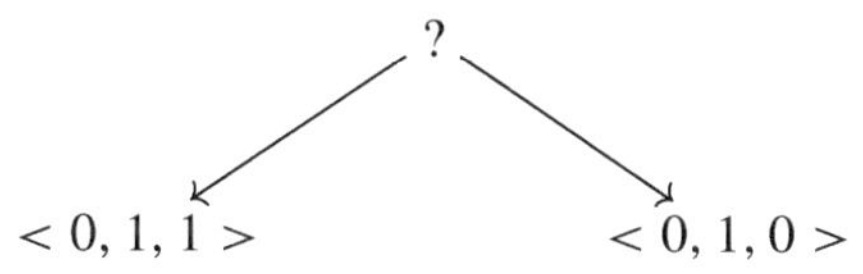

Correct Subsets of $\mathcal{C}$

Within $\mathcal{C}$, we further discern a set δ, called a *correct* set of conditions, such that:

1. $(\pi_2 \in \delta, \pi_1 \subset \pi_2) \Rightarrow \pi_1 \in \delta$;
2. $(\pi_1 \in \delta, \pi_2 \in \delta) \Rightarrow \exists\, \pi_3 \in \delta$ such that $\pi_1 \subset \pi_3, \pi_2 \subset \pi_3$.

This definition is reminiscent of that of an ideal in ring theory—in our case an element of a correct set δ is such that it forces every object that is below it in the order to belong to δ.

Example 4 Following up on the previous example, let $\mathcal{C}$ be again the set of all the finite binary sequences with the order defined above. It can be checked that the subset of $\mathcal{C}$ whose elements are the finite sequences having only $1'$s as entries is a correct set (let us call such a subset δ_1). Indeed, only finite sequences of all $1'$s can be dominated by sequences of the same kind and two such sequences are clearly compatible (it is enough to take a third sequence of all 1's longer than the two chosen ones).

Once we realize that a correct set can be discerned in an unambiguous way by its defining property (i.e. sequence of all 1's), it is remarkable to notice that the interplay between the structure of $\mathcal{C}$ and that of δ generates a natural boundary between the discernible and its complement. Indeed, since any element, say π_1 in $\mathcal{C}$ must be dominated by two incompatible conditions, say π_2 and π_3, by the property 2 above for δ at least one of them, say π_2, cannot be an element of δ. Therefore π_2 does not possess the property which discerns δ!

Badiou cleverly notes that "the concept of correct set [...] is perfectly clear for an inhabitant of M" ([1], p. 366). What he means by this is that the correct set can be discerned from within M: a suitable formal language that is codable in M is powerful enough to "see" the crisp boundaries of δ within $\mathcal{C}$. Badiou continues, "What is not yet known is how to describe a correct set which would be an indiscernible part of $\mathcal{C}$, and so of the model M" ([1], p. 367). This is our next task.

Dominations

We start with an example which reconnects with those given above ($\mathcal{C}$ is again the set of all the finite binary sequences and the order is the usual one).

Example 5 The property discerning the complement of δ_1 in $\mathcal{C}$, say δ_1^c is given by: "binary sequences containing at least one 0". For any element in δ_1 there will be an element in δ_1^c which dominates it. But this offers a beautiful way to let the discernibility of δ_1 be characterized in a structural way, with no reference to the language: for any element of δ_1, for instance $< 1, 1, 1 >$ there is at least one element in the complement, say $< 1, 1, 1, 0 >$ which dominates it. This only apparently innocent observation is at very heart of the foundation of the generic.

To formalize what we were just discussing in this example, consider a correct set δ. δ is *discernible* from an inhabitant of M if there is an explicit property λ (expressible within the model M) that distinguishes δ unambiguously:

$$\alpha \in \delta \iff \lambda(\alpha)$$

Since every condition $\pi_1 \in \delta$ is dominated by two incompatible conditions π_2 and π_3, by the second property defining δ we have that either π_2 or π_3 has to live outside δ. The importance of this remark cannot be overestimated. It is indeed upon this simple observation that we can define what a *domination* is. Let's define a *domination D* as a set of conditions such that any condition outside the domination is dominated by at least one condition inside the domination. In symbols:

$$\sim (\pi_1 \in D) \Rightarrow (\exists \pi_2)[(\pi_2 \in D)\&(\pi_1 \subset \pi_2)]$$

(here $\sim$ denotes negation).

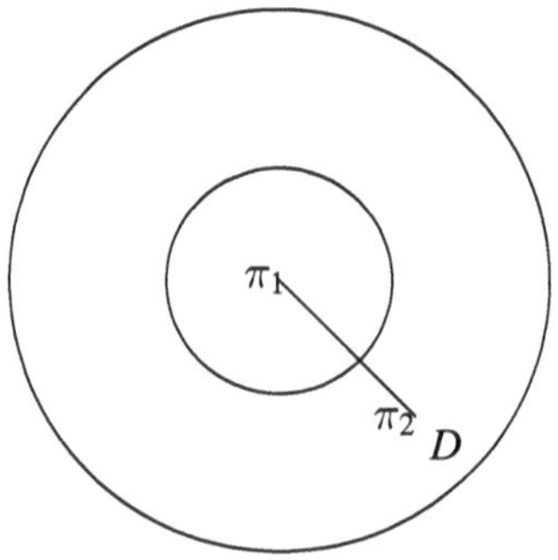

The Generic Set

We may now define a generic set. A correct set G will be said to be *generic* for M if, for any domination D which belongs to S, we have $D \cap G \neq \emptyset$. In other words, G is a correct set that has the property of having at least one element in common with all the dominations. From the point of view of M, it should be clear why G is indiscernible: "...otherwise it would not intersect the domination which corresponds to the negation of the discerning property" ([1], p. 370).

4.4.2 From M to $M[G]$

Let us remember that our goal is to find another model of ZF by *adjoining* G to M. In order to accomplish this task, we need two more intermediate steps.

Names

In spite of its indiscernibility, elements of G must be given names from within M. This is done by using a technique very close, in spirit, to transfinite induction. We begin by defining names μ of *rank 0*:

$$rank(\mu) = 0 \iff [(\gamma \in \mu \Rightarrow \gamma =< \emptyset, \pi >)]$$

where $\pi \in C$ is any condition. Essentially, names of rank 0 are thus (any possible) sets of ordered pairs, of which each ordered pair simply indexes a particular condition (an element of C) to the empty set.

Inductively we can then define names of rank greater than 0:

$$rank(\mu) = \alpha \iff (\gamma \in \mu \Rightarrow \gamma =< \mu_1, \pi >)] \ \& \ rank(\mu_1) < \alpha.$$

In effect, this defines an inductive hierarchy of names in which some name at a given ordinal rank greater than 0 is itself constituted as a set of ordered pairs, of which each ordered pair indexes a particular condition to some name as determined at a lower rank. These names, whose ordinal induction may be entirely defined from within M but which cannot in fact all be themselves elements of M, will be used to produce the "excess" of $M[G]$ over and above M by assigning a specific value to each name according to a second inductive procedure.

Reference Values

Once we can name elements of G, we want to assign *referential values* to these names which will in effect determine the essential features of the adjunction of G to M through the action of G (or rather, the strictly formal definition of G from within M as outlined above in Sect. 3.1) upon the ordinal hierarchy of names. In this way, the names will be used to indicate the generic set G sufficiently without thereby discerning or fully determining it. We start by defining referential values for names of rank 0: For names of rank 0 (which are composed of pairs $<\emptyset, \pi>$) we posit:

- $R_G(\mu) = \{\emptyset\} \iff \exists <\emptyset, \pi> \in \mu \mid \pi \in G$
- $R_G(\mu) = \emptyset$ otherwise.

Inductively, let us suppose that the referential value of the names has been defined for any rank less than α. Then, if μ_1 is a name such that $rank(\mu_1) = \alpha$, we have that:

$$R_G(\mu_1) = \{R_G(\mu_2) \mid (\exists\pi)(<\mu_2, \pi> \in \mu_1 \ \& \ \pi \in G)\}$$

Each of the names is thus formally assigned a unique value that depends solely upon which conditions are indeed elements of G in any given case, although G itself remains relatively undetermined (save for its genericity).

Adjoining the Generic Set to the Ground Model

At this point we are in a position to define $M[G]$:

$$M[G] = \{R_G(\mu) \mid \mu \in M\}$$

In plain words, $M[G]$ is the set of all the G-referential values of names that are themselves elements of M. We will not go into further technical details, but it is worthwhile, for the sake of clarity, to mention that the following, fundamental facts can be proven:

- $M \subset M[G]$, that is, $M[G]$ is a non-trivial extension of M
- $G \in M[G]$ and, moreover, G itself can be named from the point of view of M.

4.4.3 Diagrammatic Recapitulation

The diagram below summarizes the order and connections between the various steps of the construction of $M[G]$. The two triangles represent determinations involving more than one of the previous results.

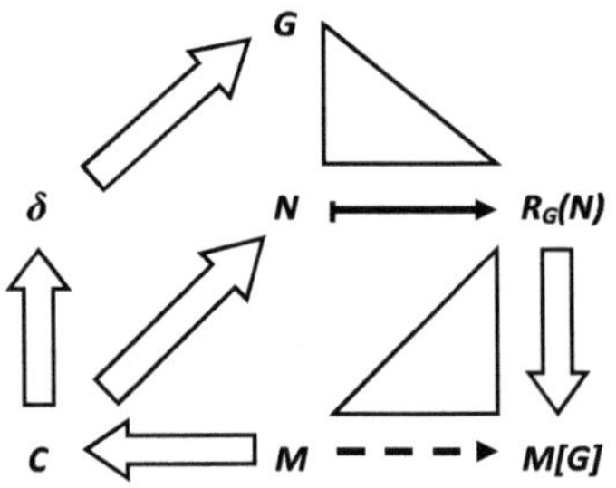

4.5 Modeling the Forcing Relation in *EG*

Cohen's mathematical technique of forcing produces a correspondence between a well defined forcing relation in M and possible truth values of statements in $M[G]$. We represent this correspondence iconically in Peirce's EG as a lifting of EG_β statements on a single sheet of assertion into a modal sheet of EG_γ. In our diagrammatic construction, discrete non-modal statements at the level of the EG_β sheet of assertion take as their referents the emergent continuous domain of $M[G]$ as expressed in modal terms in EG_γ.

4.5.1 The Quasi-Implicational Structure of Forcing and Peirce's Existential Graphs

On the one hand Cohen, in his exposition of forcing, highlights its modal nature, noting that ([3], p. 111)

> Clearly there are some properties of S which no reasonable procedure could interpret as being true or false for a 'generic' set G...Given π we will then ask, if under some procedure to be given, it is reasonable to expect that π forces a statement S about G to hold or forces $\sim S$ to hold, or whether the condition π does not force S one way or the other. *Although forcing will be related to the notion of implication it will differ from it in that given that π forces S it will not be true that any G that satisfies π will also satisfy S. What will be true is that any generic G satisfying π will also satisfy S* [notation altered to conform with the presentation above and emphasis added].

On the other hand, in his system of Existential Graphs, Peirce distinguishes between the single sheet of assertion at the level of α and β graphs and a "book of separate sheets, tacked together at points" at the γ-level, where modality emerges from relations of continuity and discontinuity between the sheets.

> In the gamma part of the subject all the old kinds of signs take new forms...Thus in place of a sheet of assertions, we have a book of separate sheets, tacked together at points, if not otherwise connected. For our alpha sheet, as a whole, represents simply a universe of existent individuals, and the different parts of the sheet represent facts or true assertion concerning that universe. At the cuts we pass into other areas, areas of conceived propositions which are not realized. In these areas there may be cuts where we pass into worlds which, in the imaginary worlds of the outer cuts, are themselves represented to be imaginary and false, but which may, for all that, be true, and therefore continuous with the sheet of assertion itself, although this is uncertain [cited in [4]].

Forcing is intrinsically relational—it distributes a determination or "control" across two heterogeneous domains. $M[G]$ is built up solely from resources available in M, but the very genericity of G guarantees that the extended model $M[G]$ is indiscernible from the standpoint of M and thus radically different or "Other". This is why the definition of G on the basis of the dominations is so important—dominations effectively formalize an intuitively epistemological or semiotic concept, namely the discernibility of a "property" of some given correct set. It is the power of the generic to name the indiscernible [1] that corresponds to the creative moment that Peirce emphasizes is central in the logical operation of abduction.

The second crucial element of the heterogeneity of M and $M[G]$ is the pair of transfinite inductions that determine the hierarchy of names on the one hand and the reference-values on the other. Notice that the names are defined independently of G while the reference-values are determined precisely by coordinating names with elements of G. If the names function first of all as indices of conditions and then, through the succession of ordinal ranks, indices of possible collections of these indices as paired with conditions, and so on, the reference-values function as the well-defined link that connects the names in M to elements of $M[G]$.

Forcing is, quite simply, a carefully-defined relation between conditions and statements in M. What matters of course is precisely how this relation is defined: the effort of building up the extended model $M[G]$ in the way outlined above becomes worthwhile solely because it enables this definition itself. Forcing is a relation in M between a given condition and a statement about the names that holds if and only if the corresponding statement about the reference-values of those names is verifiable in $M[G]$ and the given condition is an element of G.

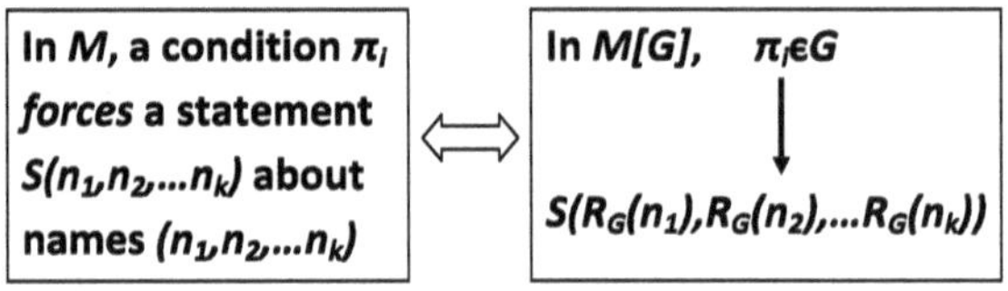

Thus, a relation that is wholly determined within M is able to signify the verifiability of statements about the extended model $M[G]$, depending upon whether or not particular conditions in M are also elements of G. Because of this, it becomes entirely determinable within M given a statement S whether or not the condition of the empty set forces S, whether or not some but not all conditions force S, and whether or not no conditions whatsoever force S. Indeed it is easily shown that these three cases or possibilities are both mutually exclusive and exhaustive. To each of the three cases corresponds a specific kind of knowledge about the statement S which takes the reference-values of the names as its argument, and hence is a statement "about" $M[G]$. The following table lays out the relevant correspondences:

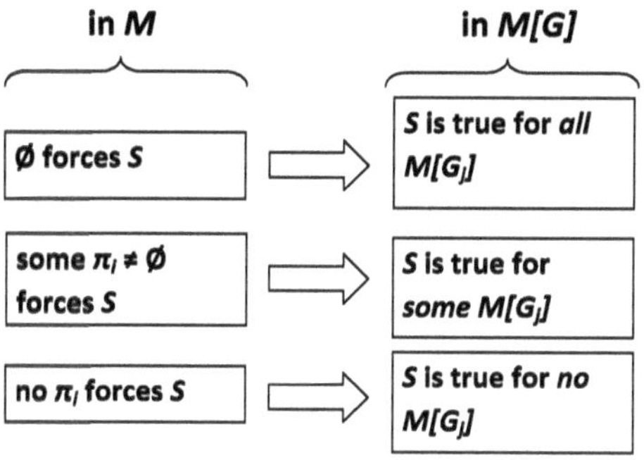

As this table shows, forcing is thus both an explicit relation in M (between conditions and statements) and an implicit correlation between M and $M[G]$ (determining at least some knowledge of the latter from the standpoint of the former, namely whether the corresponding statement about $M[G]$ is *always*, *sometimes but not always* or *never* verifiable).

4.5.2 *Diagramming Forcing as the Abductive Emergence of EG_γ from EG_β*

Within the newly constituted horizon of statements about $M[G]$ from the standpoint of M, the three "cases" may therefore be interpreted in terms of the three logical modalities expressible in Peirce's EG_γ: necessity, contingency and impossibility. Peirce's EG_γ notation is as follows:

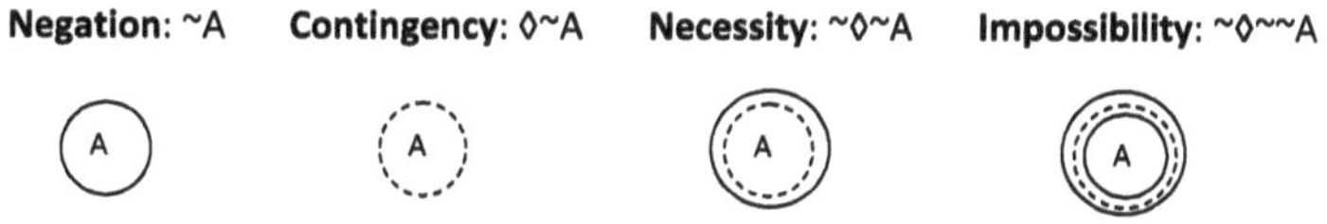

An essential aspect of abduction is that its conclusions, like those of induction, are at best merely probable. Yet even more strongly, Peirce insists that abduction is not even determinately probable—this is due directly to its "creative" element. It is exactly this character of abduction that has eluded formalization.

Yet the modalities expressible in EG_γ reflect the bare, minimal knowledge that is both necessary and sufficient for abduction. Thus if we are able to express formally the relation between non-modal EG_β statements in M and modal EG_γ statements in $M[G]$, then we will have within Peirce's graphical system a diagrammatic representation of the specific correlational structure of forcing insofar as it defines modal truth-values in $M[G]$ from within M.

The following diagram represents, in EG, the lifting of the forcing relations in the ground model M into the modal truths of $M[G]$ (for the sake of concise representation, n_k here represents the sequence $n_1, n_2, \ldots, n_k$):

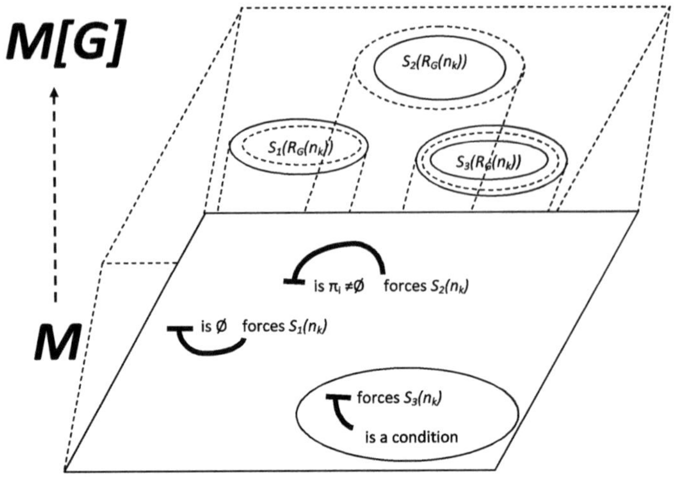

This representation shows how the set-theoretical statements and relations of forcing may be translated into the formal-iconic language of EG. This form itself is productive of new content: the EG_γ representation of forcing is itself a diagram whose internal relations carry additional implications and consequences. In particular, it illustrates the abductive character of forcing from an essentially topological perspective. In this way, we produce a tentative framework for the further investigation of the problem of a logic of abduction in terms of the problematic of the topological relations between the continuum and its possible coverings. The interested reader should also consult [5] to see how Cohen's method of forcing may be reconstituted in terms of Boolean valued models. This illuminating approach stands a sort of "middle way" between Cohen's own presentation and the sheaf-theoretical modeling of forcing in topoi as pioneered by Moerdijk in [6].

4.5.3 *Truth and the Mathematical Subject*

Our formalization of the Peircean abductive structure of forcing extends the philosophical interpretation of Cohen's work offered by Badiou in [1] while remaining fully within the formal language of EG. Badiou understands the universe of Gödel's constructible sets to correspond to being *qua* being whereas the models of ZFC that adjoin generic sets in accordance with Cohen's technique indicate a domain in which infinite procedures of truth may exceed Being as such. As Badiou himself points out, such "truth procedures" rely upon the formation of hypotheses that are always necessarily in excess of what has already been given, but are expressed in terms of the "finite fragment made up of the present state of the enquiries" ([1], p. 400):

> ...the subject solely controls – because it is such – the finite fragment made up of the present state of the enquiries. All the rest is a matter of confidence, or of knowing belief.

> Is this sufficient for the legitimate formulation of a hypothesis of connection between what a *truth* presents and the *veracity* of a statement that bears upon the names of a subject-language? Doesn't the infinite incompletion of a truth prevent any possible evaluation, *inside* the situation, of the veracity to-come of a statement whose referential universe is suspended from the chance, itself to-come, of encounters, and thus of enquiries?

Cohen's proof demonstrates mathematically that the answer to this question is no. The forcing relationship in M allows us to make definite, but limited, assertions about truths in $M[G]$. For Badiou, this relationship between the discretely statable and the generically indiscernible instantiates what he calls the *fundamental law of the subject*:

> if a statement of the subject-language is such that it will have been veridical for a situation in which a truth has occurred, this is because *a* term of the situation exists which both belongs to that truth (belongs to the generic part which *is* that truth) and maintains a particular relation with the names at stake in the statement.

Our diagram formalizes the relationship in question as the emergence in Peirce's EG_β of a continuous sheet in excess of the sheet of assertion which corresponds to the reference-domain of $M[G]$.

In his seminal paper, Louis Kauffman [7] establishes deep connections between Peirce's EG and infinitesimals from the topological perspective that naturally emerges from Peirce's ideas. In a similar spirit, we hope that further investigations into the mutual translatability of Peirce's graphs (especially the gamma graphs) with Badiou's work in set-theoretical ontology can contribute to opening new ground for the construction and investigation of a formal topological theory of continuous modalities and abductive processes. Some inroads on this research program have already been made by Zalamea in particular. The reader should examine [8].

4.6 Mathematical Ontology, Realism and Naturalism

Gironi [9] has examined Badiou's mathematical ontology from a twofold perspective that coordinates the viewpoint of structural realism and the position of epistemological naturalism.

The type of realism endorsed by Gironi is the Ontological Structural Realism developed by theorists such as French [10] and Ladyman [11]. A standard distinction in the literature holds separate the view of Ontological Structural Realism from the weaker view of Epistemic Structural Realism essentially as follows:

- Ontological Structural Realism (OSR): The ultimate constituents of reality are structures. These fundamental ontological elements are not only representable mathematically, they themselves *are* in a real sense the formal structures (for example, group-theoretical) that mathematics investigates. Reality is fundamentally structure.

- Epistemic Structural Realism (ESR): What can be known of reality extends exactly as far as and no farther than whatever mathematical structures are in fact instantiated by reality. These structures themselves may be expressed naturally by theories formulated as mathematical groups, categories, etc. Theories are structural in essence and therefore potentially realist in this epistemic sense.

A good introduction to the broader issues surrounding the Structural Realism program may be found in [12]. It should be clear in any case that OSR involves a significantly stronger theoretical commitment than ESR and that, arguably at least, only the former position truly merits the designation of "realism" in a robust sense. Recasting the issue in the terms of Badiou's philosophical project, Gironi maintains that the methodological demands of realism require at least the following ([9], p. 132):

> to recast the notion of event as triggering a rational (and defeasible) procedure of abductive discovery, as opposed to a subjective procedure of constitution, the success of which is guaranteed by sheer fidelity. The defeasibility is entailed by the non-monotonic nature of abductive inferences: in Badiouian terms, we must be open to the possibility that the result of a future inquiry will invalidate our truth procedure.

In effect, Gironi simply points out that if Badiou is to be interpreted in a realist framework, then the abductive character of Cohen's forcing must be understood primarily in epistemic rather than ontological terms. As Gironi puts it "it is necessary then to deflate Badiou's notion of an *ontology*–shattering event, reinterpreting it as the catalyst for an *epistemic* procedure puncturing the situation of *knowledge*" (ibid.). In other words, even if we accept Badiou's identification of set-theoretical mathematics with ontological *possibility* (the adequate discursive framework for representing being *qua* being), the surprising capacity of that framework to support the modal differentiation of Cohen's forcing should not be understood as ontologically but rather epistemologically relevant. The formal representation of "truth-procedures" models epistemic trajectories *relative* to real, albeit unknown, states. If, to the contrary, Badiou's own thoroughly ontological interpretation is accepted, then the core notion of realism—that reality is independent of and thus capable of surprising cognitional agents—is abdicated in favor of a pure idealism.

Importantly, then, Gironi shows how the idea of a mathematical or, better, structural ontology is capable of supporting both realist and idealist interpretations. To decide abductively for mathematical structure (that is, iconicity) as fundamental epistemic terrain is not to decide in advance either for or against realism. The question remains open, and may in fact be subject to experimental verification (as we will try to show in chapter six with reference to the mathematics underlying quantum theory). In this way, the framework of realism becomes coordinated in principle at least with the notion of epistemological naturalism. Naturalism conceives of knowledge and more generally cognition as a feature of the natural world on a par with other natural phenomena such as those studied by the sciences of physics or chemistry. It is thus intrinsically opposed to the kinds of philosophical dualisms represented by such canonical figures as Plato, Aristotle, Descartes and Kant. Whatever the difficulties confronting a robust naturalist account of knowledge in its own right, what is especially interesting in Gironi's position is the coordination of this type of view with that of structural realism. By coordinating the two positions, their mutual compatibility can be convincingly exhibited and this in itself may serve moreover as tentative support for them.

Gironi writes ([9], p. 132):

> On the one hand, the entirety of our mathematical *concepts* arises from the empirical encounter with the physical world and is propelled (on a biological level) by Darwinian evolution and the neuroplasticity of our brains, and (on a cultural level) by the socio-historical transmission of mathematical techniques. On the other, the best explanation for mathematics' success in describing the unobservable world should abductively lead us to infer that, at irreducible scales, there is nothing more to ("physical") reality than the mathematical structure used to describe it, and all such structures are ultimately depth-less iterations of the void.

The coordination of naturalism and structural realism thus appears from Gironi's point of view as a solution to the essentially Kantian problem of how mathematically exact science of the natural world is in fact possible. Rather than following the route of Kant's transcendental idealism which locates the foundations of arithmetical and geometrical structure in the a priori cognitive frameworks of (Euclidean) space and time, the path proposed here would take mathematical structure as such to be the ontologically prior ground of any possible cognitional being. Thus in place of the a priori necessity of objects of cognition conforming to subjective conditions of experience, one finds a single frame of mathematical structure enveloping both objects and subjects uniformly. Moreover, it is no longer necessary to distinguish in principle between the object of knowledge and the thing in itself, as it is in Kant, because the central thesis of structural realism is that things-in-themselves precisely *are* structures. Structure is thus both (and identically) what is and what is eminently knowable. If there is no remainder or residue of existence outside of structure, then there is no reason to posit an ontological distinction between what is in itself and that mere part or aspect of what is that is in principle intelligible or epistemically accessible. Nothing in principle separates the real from the cognition of the real, although there may of course be de facto inaccessible realities for some or all actual cognitional agents.

It is neither a question here of defending Gironi's position as it stands nor of critiquing it, but rather simply of indicating how it sheds light on certain core issues for Badiou's account of set-theoretical ontology and his use of Cohen's technique of forcing as an epistemic model. In particular it shows how the set-theoretical mathematical setting is capable of modeling "pure" structure in an immanently mathematical way, but that the types of relation that characterize experiential and concrete epistemic engagement with the actual world of nature, as in the processes of natural science, remain out of reach. In the next chapter we will examine Badiou's own attempt to address this lacuna in the sequel to *Being and Event*, the more phenomenologically-oriented *Logics of Worlds*.

4.7 Conclusion

Badiou's use of Cohen's forcing technique to model epistemic "truth-procedures" within the framework of an ontological interpretation of Zermelo–Fraenkel set theory may be understood as a large-scale abduction holding potential explanatory power with respect to a variety of traditional philosophical problems, many of which (like that of the status of the epistemic subject as such) might appear intractable outside of some such formal framework. Within this abductive context, the problem of scientific realism arises in a uniquely inflected mode suggesting that mathematical structure may be not only an epistemic tool or mere ontological model but a full-throated realist ontology that would be capable of *explaining* the evident (yet still highly problematic) link between knowledge and being. Not only then do certain deep and highly textured relations between epistemology and ontology appear in Badiou's work, perhaps more importantly the mathematical framework of abduction appears to take on a fundamental role in its own right, generative perhaps in a radical sense of the ontological-epistemic duality as such.

References

1. A. Badiou, *Being and Event* (Continuum, London, 2005)
2. K. Gödel, *The Consistency of the Continuum Hypothesis* (Princeton University Press, Princeton, 1940)
3. P. Cohen, *Set Theory and the Continuum Hypothesis* (Dover, Mineola, 2008)
4. F. Zalamea, Peirce's logic of continuity: existential graphs and non-cantorian continuum. Rev. Mod. Log. **9**(29), 115–162 (2003)
5. J. Bell, *Toposes and Local Set Theories: An Introduction* (Dover, Mineola, 2008)
6. S. Mac Lane, Moerdijk, *Sheaves in Geometry and Logic: A First Introduction to Topos Theory. (Universitext)* (Springer, New York, 1994)
7. L. Kauffman, Peirce's existential graphs. Cybern. Hum. Knowing **18**, 49–81 (2001)
8. F. Zalamea, Peirce's Continuum: A Methodological and Mathematical Approach. http://www.scribd.com/doc/16392109/Zalamea-Peirces-Continuum-Part-1
9. F. Gironi, *Naturalizing Badiou: Mathematical Ontology and Structural Realism* (Palgrave Macmillan, New York, 2015)
10. S. French, Structure as a weapon of the realist. Proc. Aristot. Soc. **106**, 167–185 (2006)
11. J. Ladyman, What is structural realsim? Stud. Hist. Philos. Sci. **29**, 409–424 (1998)
12. E.M. Landry, D.P. Rickles (eds.), *Structural Realism: Structure, Object and Causality* (Springer, Berlin, 2012)

Chapter 5
Phenomenology and Iconicity in Badiou's *Logics of Worlds*

5.1 Introduction

Is it possible to extend in an abductive manner the specifically iconic dimension of the mathematical modeling of reality implicit in Badiou's austerely binary ontology? To a certain extent, Badiou himself has already done so. *Logics of Worlds*, Badiou's sequel to *Being and Event*, represents the evolution and enriching of his ontology as still firmly rooted in Mathematics, but now including and oriented by a formal calculus of relations, that is, as developing in the direction of Category Theory, which, as we have argued throughout the present work, is the formal ground for iconicity to thrive.

Logics of Worlds represents a major effort to bring the philosophical notion of *manifest being*, or phenomenology to a new level of formal precision: in the same way as abduction lends itself very naturally towards modal and intuitionistic logical systems, Badiou's interpretation of the logic of phenomenal appearing is based on the notion that the algebra of relations it involves is intrinsically non-Boolean. Interestingly, the ultimate insights offered by this approach end up exceeding the limits in which Badiou's own investigation is initially framed. Instead of simply providing a logic of phenomenal appearing, the sheaf-structure that is used to coordinate Badiou's background ontology of "extra-worldly" sets on the one hand with a non-Boolean algebra of "intra-worldly" relations on the other in fact suggests that ontology itself might be more creatively open and abductive than Badiou's philosophical commitment to Platonism allows.

Although a very careful analysis [1] reveals some deep problems in the overall mathematical structure of his work, we highlight the fact that Badiou nonetheless achieves a genuine philosophical result, almost *malgre lui*. Instead of trying to reinterpret mathematics in a philosophical context (or vice versa), he succeeds in defining the structure of what he calls a "complete form of the onto-logy of worlds", by cleverly positioning his theory at a level of abstraction that is somewhat optimal in both a mathematical and a philosophical sense.

G. Caterina and R. Gangle, *Iconicity and Abduction*,
Studies in Applied Philosophy, Epistemology and Rational Ethics 29,
DOI 10.1007/978-3-319-44245-7_5

Rather than attempting to explicate and evaluate Badiou's ontological-phenomenological proposal in its entirety, our concern in the present context is merely to examine the unique role that mathematics plays for Badiou as a formal space for modeling certain philosophical questions. We aim to show that mathematics in fact functions in two distinct ways for Badiou and that distinguishing these two senses helps to clarify both the limits and the real stakes of Badiou's project. On the one hand, very general mathematical domains (in particular, set theory and sheaf theory) are used to model abstract conceptual regions within philosophy. An explicitly Platonic metaphysics is modeled in the mathematical "universe" of sets, and a less clearly defined Kantian–Husserlian notion of transcendental structures of manifest being, or appearance, is modeled by the relational logic of sheaves. Yet on the other hand, mathematics also serves to bridge and formally coordinate these otherwise distinct conceptual domains. The "logic of appearing" represented by Heyting algebras supporting sheaves is equally a logic of relations *between* the phenomenological and the ontological. Thus the logic of sheaves really serves two distinct albeit closely related functions in Badiou's overall system. It both models purely manifest or "worldly" relations *and* models the relations that link ontology and phenomenology intrinsically to one another and are thus at least potentially determinative of the internal structure of both domains interdependently. The question of the priority of relations over terms is not or should not be, mathematically speaking, decided here in advance.

In short, we claim that the mathematics of sheaves that Badiou uses solely to undergird his phenomenology is at least potentially a candidate for modeling and experimenting philosophically at the ontological level as well. The logic of appearing as formalized by Badiou does not necessarily remain external to and independent of the logic of being but in fact potentially calls into question via its own internal structure the assumption, maintained throughout *Being and Event*, that ontology must be governed by a Boolean and classical logic and necessarily remains indifferent to any internal relationality. The particular case of Badiou thus serves as a concrete and highly instructive example of how mathematical formalism may not only work to represent but can also guide and even challenge philosophical presumptions, even those it is intended mutely to support. Mathematical iconicity here plays a strong and genuinely abductive role.

In order to show how this is the case, we examine several of the most important mathematical constructions used by Badiou in *Logics of Worlds*. As will be shown, the mathematical framework that comes eventually to integrate these diverse components into a unified yet internally variable and highly flexible phenomenology, or "logic of appearing" is the theory of presheaves and sheaves that we have employed in previous chapters to organize the analysis of iconicity. The most important result that comes out of these reflections is that the difference between Boolean and non-Boolean Heyting algebras emphasized by Badiou in a limited way is not only a purely formal matter internal to mathematics but potentially represents and provides new insights into some of the most fundamental philosophical problems concerning the relationships among being, truth and knowledge, even more so than Badiou himself suggests.

The mathematical constructions we will examine to elucidate this result are the following:

1. Badiou's general notion of the *transcendental* of a phenomenal world as represented mathematically by a Heyting algebra.
2. The particular case of Peirce's EG_α considered as a category (discussed in chapter three) as one example of a transcendental in Badiou's sense that is canonically Boolean in structure but also generative of non-Boolean Heyting structures.
3. The Id function that measures the "degrees of appearing" or – from the present point of view – "degrees of iconicity" of the components of a given world.
4. Badiou's notion of *atoms* as well as the derivative relational notions of *localization* and *compatibility*.
5. The integral consolidation of these conceptual fragments into the mathematical framework of *sheaves*.
6. Badiou's concept of *points* as mappings of "collapse" from the rich and generally non-Boolean context of Heyting algebras into the much simpler logic of a 2-valued Boolean algebra.

After explaining the conceptual role of each of these constructions in turn, we then reflect more broadly on Badiou's proposal from the standpoint of the interlinked problematics of abductive reasoning and iconicity. We conclude by raising anew the questions of naturalism and scientific realism that served as a critical riposte to Badiou's ontology in chapter four.

5.2 Badiou's Transcendental

The overall motivation behind Badiou's "Greater Logic" (he takes this designation from Hegel) is clearly outlined in the introduction to Book 3 of *Logics of Worlds* ([2], p. 99):

> Where the earlier book followed the thread of ontology, my current undertaking, placed under the rubric of the transcendental, unravels the thread of logic. Previously, I identified situations (worlds) with their strict multiple-neutrality. I now also envisage them as the site of the being-there of beings.

The notion of the *transcendental*, denoted by T, is therefore central to his work.

Here is Badiou again, commenting on the move from *Being and Event* towards *Logics of Worlds* (ibid.):

> In *Being and Event*, I assumed the dissemination of the indifferent multiple as the ground of all that there is, and consequently affirmed the ontological non-being of relation. Without going back on this judgment, I now show that being-there as appearing-in-a-world has a relational consistency. I have established that "mathematics" and "being" are one and the same thing once we submit ourselves, as every philosophy must, to the axiom of Parmenides: it is the same to think and to be. It is now a matter of showing that "logic" and "appearing" are also one and the same thing. "Transcendental" names the crucial operators of this second identity.

The idea here is very simple, consisting in taking a completely pragmatic standing from the point of view of worldly appearance, or phenomenality, in which primacy is given to relations between objects rather than to the ontological status of the objects themselves.

In order to unfold and grasp the complexity of a relational world, Badiou builds up a theory of internally structured *phenomenological ranks* based on the fact that any possible form of relational complexity is nothing else but the reflection of a highly structured mathematical object (which he calls the *transcendental* T), an object that in mathematics corresponds to the notion of a Heyting algebra. It is important to notice that, although Badiou's ultimate goal is to construct a purely relation-based model of ontological manifestation, his effort is rooted in the algebraic system employed to represent the world, and therefore we begin by outlining his notion of the transcendental from a purely algebraic point of view.

5.2.1 *T as a Heyting Algebra*

Badiou defines T stipulatively as a complete distributive lattice, that is, a set endowed with the following operations, relations and properties.

1. An order relation $\leq$ defined on the elements of T.
   More precisely, this means that given any two elements x, y in T, we have that

   (a) $x \leq x$ (reflexivity)
   (b) $(x \leq y \ and \ y \leq z) \implies x \leq z$ (transitivity)
   (c) $(x \leq y \ and \ y \leq x) \implies x = y$ (anti-symmetry)

   It should be noted that it is not the case in general that, in a set S endowed with an order relation of this kind, all the elements need to be "linked" by $\leq$: it could be the case, for instance, that there are x and y in T such that neither $x \leq y$ nor $y \leq y$. This is the reason why a set endowed with such an order relation is called a *partially ordered set*. Nonetheless, a globally synthetic structure coordinating all the elements of T into a coherent whole *is* in fact imposed by the remaining conditions enumerated below, which serve in this respect very much like the axioms of a theory.
2. A least element μ in T.
   μ is an element of T such that $\mu \leq x$ for all $x \in T$, that is, μ is the "bottom" element (if we think of the order moving in an upward direction). It is an easy exercise to show that μ is unique.
3. An operation of conjunction $x \cap y$ for any two elements x, y of T.
   $x \cap y$ is the greatest lower bound with respect to x and y. This means, that, if we interpret $\leq$ as meaning "less than or equal to", then $x \cap y$ is, by the definition, the greatest amongst all the elements of T which are smaller than both x and y.
4. The existence of an *envelope* ΣB, for any subset $B \subset T$.
   ΣB is the least upper bound with respect to all the elements of B. This means

that ΣB is the element which is the smallest amongst all the the elements of A which are larger than or equal to all the elements of B. This is the dualized, multi-element notion of the conjunction.

5. Distributivity of $\cap$ over Σ.
The conjunction $\cap$ distributes over the envelope Σ. That is, for any $B \subset T$ and any $d \in T$, the following holds:

$$d \cap \Sigma B \ = \ \Sigma\{d \cap x \mid x \in B\}$$

Since, ultimately, T will play the role of the set of truth-values for certain types of evaluations defined on a world/set A into T – as we will see in the next section – in what follows we reformulate two well-known classical logical operations within the algebraic structure defined above.

First of all we define *negation*. We then establish a *dependence relation* corresponding roughly to logical implication.

1. *Negation*: for any $p \in T, \neg p \ = \ \Sigma\{q \mid p \cap q = \mu\}$
It can be shown that $p \cap \neg p = \mu$. This means that the conjunction of a transcendental degree and its negation is always equal to the minimum.
Another property of $\neg$ in the context of Heyting algebras is that, for any $p \in T$ the following holds:

$$\neg\neg p \geq p$$

This latter relation is really what sets apart Heyting algebras from classical (Boolean) logic where, instead, the following law of double negation holds:

$$\neg\neg p = p$$

Finally, it can be shown that, in any Heyting algebra, the negation of the minimum coincides with its maximal element:

$$\neg\mu \geq p \ \forall p \in T$$

We can then denote $\neg\mu$ by M.

2. *Dependence of q with respect to p*: for any $p, q \in T$, $p \Rightarrow q \ = \ \Sigma\{t \mid p \cap t \leq q\}$
The meaning of this latter operation is the following: consider the set S of all the elements $t \in T$ such that, when conjoined with p, return an element less than q and take the envelope of S. For instance, if p and q are such that $p \leq q$, then it can be shown that $p \implies q = M$: the degree of dependence is maximal in this case. On the other hand, if p and q are such that $p \cap q = \mu$, we have that $p \implies q = \mu$ as well: the degree of dependence is minimal in this case.

More generally, if we regard T as a structured set of truth values, it can be shown that certain fundamental relations that hold in classical logic do not hold here.

In particular, the following do not necessarily hold in T:

1. the law of double negation, or $\neg\neg p = p$
2. the principle of the excluded middle, or $p \cup \neg p = M$
3. the fact that the equation $\neg p = \mu$ has only one solution for p, which is M

If any of the above relations does in fact hold, T is said to be a *Boolean* algebra, and it will be denoted by T_0 to highlight this crisper nature.

Example 6 The typical example of a Boolean algebra is the set $B = \{0, 1\}$ with the natural order $0 \leq 1$ and with $\cap$ corresponding to the logical operator AND.

Example 7 An important class of Heyting algebras that, in general, are not Boolean, is constituted by the lattice of open sets of a topology, along with operations of arbitrary unions $\cup$, finite intersections $\cap$, $\mu = \emptyset$ and $a \Rightarrow b = int(\overline{a} \cap b)$, where the overline indicates the set-theoretical complement and $int(x)$ the interior of x (the largest open set included in x).

5.3 EG_α as a Concrete Model of the Transcendental

The purely algebraic description of the transcendental might appear to be prohibitively abstract. For heuristic purposes, it is useful to show how Peirce's *alpha* level of Existential Graphs, denoted by the symbol EG_α, when conceived as the category $\mathcal{G}$ (see the analysis in chapter three), provides a concrete model of Badiou's notion of the transcendental of a world and in this way helps to illustrate the concept. Indeed, the pragmatic perspective conveyed by a categorical interpretation of EG_α as generating iconicity at a second diagrammatic level immediately opens up a whole new universe of abductive opportunities.

In particular, we can move beyond the information carried by one single graph and start looking at the structure of organized subsets of $\mathcal{G}$. There are two main reasons why we believe that this is an important shift in perspective. In the first place, in agreement with the self-interpretative nature of graphs, it is desirable to think of $\mathcal{G}$ as a lattice of truth-values: although Brady and Trimble in [3] already made this connection implicitly by proving the correspondence of EG_α with Boolean algebras, we show that there is a natural topology on EG_α leading to an extension to algebras that are more general than Boolean – in fact, Heyting algebras. As a result of this process, we can outline abductive inference in a much more "flexible" way, leaving space for the "preservation of ignorance" [4] which does not seem to be grasped by locking abduction into an algorithmic framework.

Below, we first present an alternative definition of Heyting algebras somewhat more suitable to be employed within the concrete model provided by EG_α. The reader should check how this characterization is equivalent to the one given above.

Let S be a set endowed with a partial order $\leq$, three binary operations $\wedge$, $\vee$ and $\rightarrow$ and two special elements $\top$ and $\bot$. S is then a *Heyting algebra* if the following hold for all g, h and k, elements of S:

1. $g \leq \top$
2. $g \wedge h \leq h$
3. $g \wedge h \leq g$
4. $g \leq h$ and $g \leq k$ implies $g \leq h \wedge k$
5. $\bot \leq g$
6. $g \leq g \vee h$
7. $h \leq g \vee h$
8. $h \leq g$ and $k \leq g$ implies $h \vee k \leq g$
9. $g \leq (h \rightarrow k)$ iff $g \wedge h \leq k$

We also define the *complement* of g as $\neg g = (g \rightarrow \bot)$. Heyting Algebras are extensively used in Intuitionistic Logic because they do not require the law of the excluded middle to hold. That is, in general, $\neg(\neg g)$ need not be equal to g.

It can be easily shown that the category $\mathcal{G}$ as constructed in chapter three is a Heyting algebra. All of the above operations and relative conditions can be translated into EG_α as interpreted within $\mathcal{G}$. In particular, the operations $\wedge$, $\vee$, $\Rightarrow$ are expressed by functors $\mathbf{G} \times \mathbf{G} \longrightarrow \mathbf{G}$ where $\mathbf{G}$ is, as in chapter three, the discrete category (or set) of all well-formed alpha graphs. These functors are illustrated iconically below:

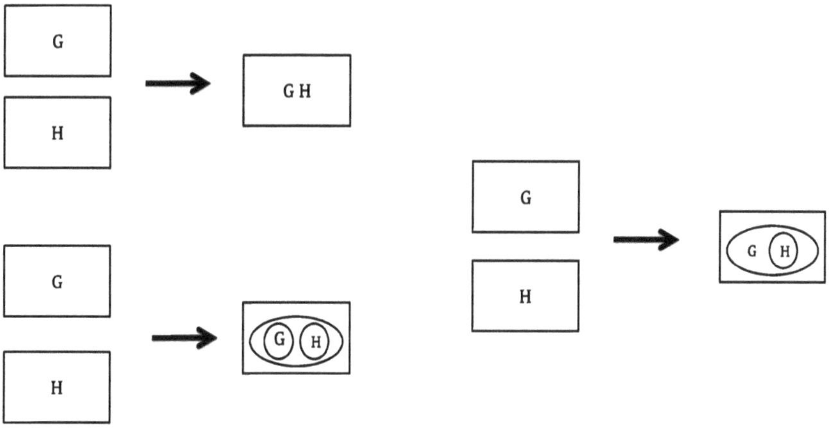

That $\mathcal{G}$ (or a suitably collapsed $\mathcal{G}$) meets condition 4, for instance, may be shown easily. Every graph G may be "doubled" to form the graph GG (this itself is demonstrated through application of write double cut, iteration, erase double cut). Then using Theorem 1 from chapter three twice transforms GG to HK. The diagrams below present the intuitive idea (the dotted arrow represents what needs to be "derived" from the two given arrows taken together).

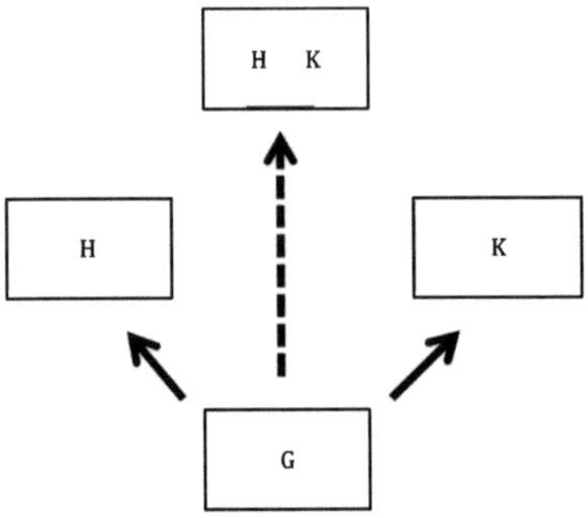

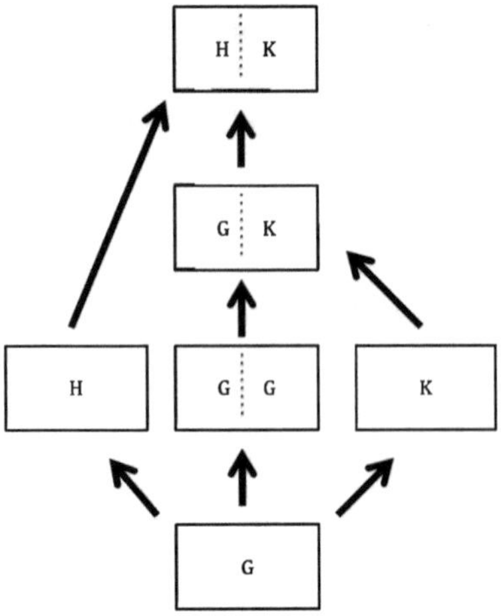

It is an easy exercise to show that the conditions (1) through (9) hold. In a sense, this is a trivial point because it has been shown that $\mathcal{G}$ is equivalent to the theory of Boolean Algebras [3] and Boolean Algebras are a subset of Heyting Algebras. What is interesting is that the conditions for being a Heyting algebra may be expressed by functors with an immediate iconic representation of the algebraic structure of $\mathcal{G}$.

Everything we have done so far has taken place in a Boolean context. Yet having translated the graphs into a categorical framework opens up a new set of perspectives that might not have been otherwise evident. This is precisely the naturally abductive character of iconic signs [5]. In particular, we can extend our notion of "object" from single graphs to structured subsets of $\mathcal{G}$. As we will show below, this simple move corresponds to a well-established notion of open sets of a topological space, which has been extensively used by logicians to create models of truth alternate to the Boolean ones.

Let us first define the *upper sets* of $\mathcal{G}$.

For any $g \in \mathcal{G}$ we define $g \uparrow$ as

$$g \uparrow = \{h \in \mathcal{G} \; g \rightarrow h\},$$

that is, intuitively, $g \uparrow$ represents the collection "of the futures" of g. We now declare a subset U of G an *upper closed set* if $g \uparrow \subseteq U$ for any $g \in U$.

Lemma 1 *The set $UC(\mathcal{G})$ of all upper closed subsets of $\mathcal{G}$ is a topological space.*

Proof The denumerable union of upper closed sets is an upper closed set and the finite intersection of upper closed sets is an upper closed set. Moreover, the empty set and $\mathcal{G}$ are also upper closed sets.

Thus, Peirce's system of Existential Graphs at the alpha level serves as a particularly rich instance of the kind of structure that Badiou wishes to use to measure degrees of existence and appearance *relative to the relationality of a given world*. The fact that Peirce's EG_α is general enough to model all of classical propositional logic *and* at a second level opens up onto a different, non-classical logic helps to show how the idea of a transcendental in Badiou's sense can possess a natural variability that is not wholly arbitrary.

5.4 Degrees of Identity and Appearing

How is a transcendental *used?* For Badiou, it serves to define and synthesize the various levels of relation and existential appearing of objects in a given world.

The first move towards this thoroughly relational phenomenology is that of defining a mapping which measures the degree of identity between any two elements of a given world (here by *world* we mean an arbitrarily given set A). The properties imposed on this function, along with the algebraic structure, will pave the way to Badiou's "complete form of the ontology of worlds."

5.4.1 The Function Id

We define *a function of appearing* $Id(x, y) : A \times A \to T$ with the following properties:

1. $Id(x, y) = Id(y, x)$
2. $Id(x, y) \cap Id(y, z) \leq Id(x, z)$

1. is equivalent to the fact that ([2], p. 573)

> [...] the transcendental degree which measures, in a given world, the identity of one apparent to another, also measures the identity of this other apparent to the first: the function of transcendental indexing is symmetrical. (*Logics of Worlds*, Statement 28).

2. says that ([2], p. 573)

> [...] the intensity of co-appearance, or conjunction, in a given world, of the identity of one apparent to another, and then of this other apparent to a third, cannot surpass the degree of identity that can be directly evaluated between the first and the third. With regard to conjunction, transcendental indexing obeys a condition of triangular inequality. (*Logics of Worlds*, Statement 29).

Given that a world (or a part of it) for Badiou is just an unstructured set, the function of appearing Id has the effect of adding an order relation to any pair of elements of A: if $Id(x, y) \leq Id(x, z)$ we can think of z to be "more like" x than y.

This common-sense, realist interpretation seems to break down when we consider its diagonal $E(X)$:

$$E(x) = Id(x, x).$$

We will refer to $E(x)$ as the degree of existence of x and it is immediately clear that, contrary to an intuitive or common-sense perspective, degrees of existence are not always globally maximal: it may see strange to claim that x looks like herself more than y does (that is, $E(y) \leq E(x)$).

The point here can be clarified by the following relation between Id and E, that follows directly by 1. and 2.

$$Id(x, y) \leq E(x).$$

The latter says that ([2], p. 573)

> [...] an apparent in a world cannot exist in that world less than the extent to which it is identical to another apparent.(*Logics of Worlds*, Statement 30)

What the previous result entails is the fact that the degree of existence $E(x)$ of $x \in A$ is maximal only with respect to the degree of identity with everything else in A, whereas, in general, it should not be surprising that there exists y with a degree of existence $E(y)$ greater or less than $E(x)$. In other words, degrees of existence are indexed by elements of the transcendental and can be thought as *degrees of iconicity*: $E(x) \leq E(y)$ may reflect the fact that x is a lesser iconic representation than y (with respect to their relation with their respective objects).

Finally, we notice that any element $a \in A$ defines canonically a function

$$a(x) : x \to Id(a, x)$$

This amounts to saying that there are certain ways of defining the degree of identity map Id that are rooted in the "real", that is, they are represented by an element of the world.

This latter definition is crucial to the development of the rest of this chapter, and it can be thought of as the first step towards more general representation results, such as Yoneda's lemma and Yoneda's embedding that we will discuss more fully in the following chapter. It is a construction that illustrates the richness of relational structures in a very immediate way: the function $a(x)$ generated by any element a can be regarded as a sort of perspective on the category to which a belongs "from the point of view of a". $a(x)$ encodes all the possible relations between a and an arbitrary element y, and in turn it generates a family of contravariant functors that can be patched together to obtain a genuine set-theoretical representation of the entire category. The basic construction we must elaborate at this stage, however, is what Badiou calls, somewhat provocatively, the *atoms* of his phenomenology.

5.5 Atoms

For Badiou, an *atom* is a constraint on how the pure sets of ontology may appear in a phenomenal world. The degree of existence function $E(x)$ is induced by Id and constrained by the axioms imposed on the latter. In general, however, we can think of associating elements of T to elements of the world without any constraint at all, defining arbitrary functions from A to T. In this section we look at the restrictions Badiou imposes on these correspondences in order to make his sheaf-theoretical machinery work.

Below we present three ways in which "degrees of existence of an element of a world" can manifest themselves, ordered by their level of generality.

1. First, and most generally, we will call an *object-component* any function

$$\pi(x) : A \to T.$$

2. Secondly, we will call an *atom* $\alpha(x)$ an object-component with at most a unique maximal element, that is, with the property that there is at most only one $x_0 \in A$ such that $\alpha(x_0) = M$.
3. Finally, we say that an atom is *real* if it can be represented by an element $a \in A$, that is, if there is $a \in A$ such that $\alpha(x) = a(x)$.

Here below is Badiou's comment on 2. ([2], p. 248):

> We call atomic object-component, or simply "atom", an object-component which, intuitively, has at most one element in the following sense: if there is an element of A about which it can be said that it belongs absolutely to the component, there is only one. This means that every other element that belongs to the component absolutely is identical, within appearing, to the first (the function of appearing has the maximum value M when it evaluates the identity of the two elements in question).

This comment, meant to clarify the nature of an atomic-component, highlights the non-realist framework in which Badiou is operating. He is indeed redefining the ontological status of equality: two elements x and y are equal, or the same (indistinguishable), if their degrees of existence are both maximal, that is if $E(x) = E(y) = M$. We then say that x_0 belongs absolutely to a given component $a(x)$ if its value via $a(x)$ is maximal, that is if $a(x_0) = M$. Therefore an atom is simply a function $A \longrightarrow T$ which synthesizes maximality both in a local (with respect to $a(x)$) and global (uniqueness of the maximum M) sense.

Going back to the opening question of this subsection, we list below the restriction that an object-component needs to satisfy in order to be an atomic component.

It can be shown that if an object-component $\alpha(x)$ satisfies the following properties-axioms, then it is an atom.

1. $\alpha(x) \cap Id(x, y) \leq \alpha(y)$
2. $\alpha(x) \cap \alpha(y) \leq Id(x, y)$

Property 2 expresses the uniqueness – modulo the remarks made above – of an element belonging to the atom absolutely, whereas 1 constrains an element y to belong "strongly" to the component of an atom α if x also belongs strongly to the same component and, at the same time, x is also "very identical" to y. We are now prepared to understand Badiou's unusual notion of *materialism* in this formal context.

5.5.1 The Postulate of Materialism and the Object

The postulate of materialism, as formulated by Badiou, is that

every atom is real

It follows that, in Badiou's own words, the ([2], p. 222)

> [...] logic of atoms of appearing ultimately concerns a certain type of correlation between the elements of A and the transcendental degrees, which are themselves elements of the transcendental T.

Here Badiou makes the most radical of the moves in order to carry out his philosophical project: single handedly, he filters out all of the object-components (including the atomic ones) which are not real (in the sense specified above). In mathematical terms, this is equivalent to restricting these to functions that (1) are atomic in the sense described above (at most one maximal element) and (2) are represented by an actual object in A ([2], p. 251):

> In effect, [this] stipulates that the virtuality of an apparent's appearing in such and such a world is always rooted in its actual ontological composition.

The postulate of materialism sets the ground for the emergence of the concept of object ([2], p. 251):

> By "object" we understand the couple formed by a multiple A and a transcendental indexing Id, a couple which is written (A, Id), under the condition that every atom of which A is the support be real; in other words, that every atomic component of the appearing of A be equivalent to a real atom $Id(a, x)$ prescribed by an element of A.

This is yet another manifestation of the different notion of equality when we move from an ontological to a phenomenological context: two set-theoretically distinct elements a, b of A may well be the "same" object to the extent that their induced real-atomic components $a(x)$ and $b(x)$ coincide.

The reader familiar with category theory will recognize the analogy with the notion of categorical isomorphism between objects of the same category.

5.5.2 Localization

What makes Heyting algebras so appealing as mathematical structures is that they are built upon both an algebraic and a topological ground. Badiou's insight is to make use of this mathematical duality in order to define the complementary notions of *localization* and *compatibilty* – two constructions that are fundamental to describing the synthetic nature of his ontological-phenomenological scheme.

We first define the *localization* of a real atom $a(x)$ to the element to the transcendental $p \in T$ as the function $a(x) \cap p$, which we will denote by the symbol $(a \int p)(x)$.

The best way to visualize this operation is to think of the localization as a sort of "truncation of $a(x)$ at the level p". In other words, the result of applying $a \int p$ on any element of the world x is to produce p if p is below $a(x)$. By the same token, the result of applying $a \int p$ on any element of the world x is to produce $a(x)$ if p is above $a(x)$.

The diagram belows illustrates how localization works (here $a \int p$ is evaluated at element $x_0 \in A$).

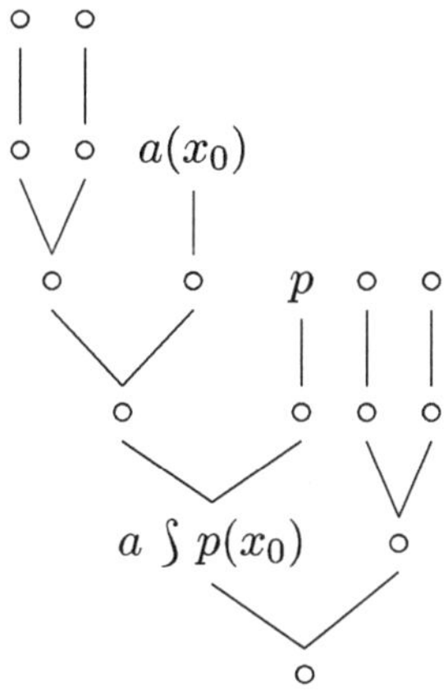

An important fact is that it can be shown ([6], p. 254) that the localization of a (real) atom to an element of the transcendental $(a \int p)(x)$ is also a real atom. Because of this, there will always be an element $b \in T$ such that $(a \int p)(x) = b(x)$. In plain words, this is equivalent to saying that "the element b is the transcendental localization of the element a, relative to the degree p".

Badiou clarifies this point further ([2], p. 254):

> We can therefore see that the global analysis of objects – of which real atoms are the foundation, at the same time that they guarantee the connection between the logic of appearing and the mathematics of being – can issue into a local analysis without losing its guiding thread. Localized by an element of the transcendental, an atom remains an atom.

What is really at stake here is the fact that the identity $a \int p = b$ entails a relation between elements of A ("ontological constitution of being") induced by the the "transcendental constitution of appearing".

In next subsection we will see how localization enables us to actualize a precise topological notion of "proximity" among the elements of an object.

5.5.3 Compatibiilty

We will say that two elements of A, say a and b, are *compatible* (relative to the transcendental T of this world, and to the transcendental indexing Id of A on T), if

$$a \int E(b) = b \int E(a)$$

We will denote this relation of compatibility between a and b with the symbol

$$a \ddagger b.$$

It should be noticed that the both $a \smallint E(b)$ and $b \smallint E(a)$ here are thought as actual functions quantified over all of A: a and b are compatible to the extent that localizing a over the degree of existence of b induces an atomic component which behaves exactly like the function induced by localizing b to the degree of existence of a.

The apparently obscure meaning of this relation is made clear when we lift it from the topological to the algebraic level. It can be indeed proved that the following are equivalent:

1. $a \ddagger b$
2. $E(a) \cap E(b) = Id(a, b)$

The diagram below illustrates this:

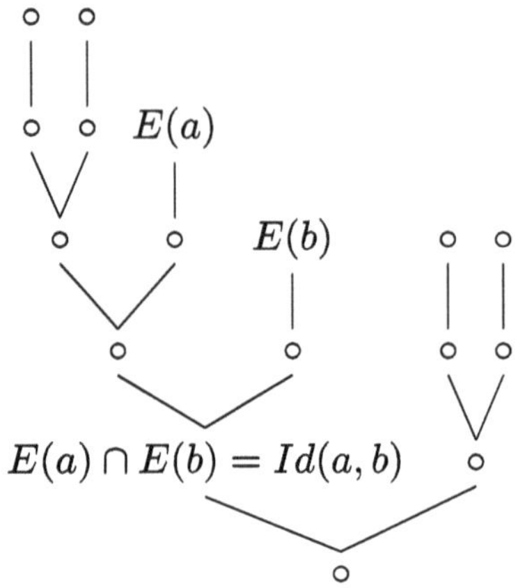

This really means that two elements a and b are compatible if "the common part" (the conjunction) of their respective degree of existence $E(a)$ and $E(b)$ is equal to "how much they look alike" (their degree of identity $Id(a, b)$).

In this sense, compatibility entails a relation of "affinity" between (compatible) elements, which – as we will show in the next sections – will play a key role in the local-to global process which will complete Badiou's ontological-phenomenological scheme.

5.6 Worlds as Sheaves

At the heart of Badiou's mathematical reframing of phenomenology rests the notion of sheaves. Thus the traditional philosophical problems concerning manifest or existential being as distinguished from ontological or essential being are formalized by

Badiou within the same framework of local-to-global and global-to-local relations that we have used throughout the present work to model the iconic character of abductive inference. It would be precipitous to conflate the two models without further ado, but it should be clear nonetheless that a significant convergence of the two otherwise distinct projects *simply at the level of form* points at least to the possibility of generating hypotheses that might translate readily from one to the other. At any rate, in order to understand exactly how Badiou employs the relevant concepts in his work, it will be helpful first to recall the formal definition of a sheaf as a specification of the more general notion of a presheaf subject to certain coherence restrictions.

5.6.1 Presheaves Over a Topological Space

The most intuitive definition of sheaves is given in the context of topological spaces.

Let X be a topological space and let $\mathcal{C}$ be a category (for instance, the category of sets). We first define a *presheaf* to be a contravariant functor F from the category of the open sets U of X (structured by inclusion) to $\mathcal{C}$, where $F(U)$ is an object of $\mathcal{C}$ and for each $V \subseteq U$ there is a correspondent morphism $res_{V,U} : F(U) \to F(V)$.

If, in particular, we consider the case when $F(U)$ is the set of continuous functions on U, to any $f \in F(U)$ is possible to associate the restriction of f to V, $f_{|V} \in F(V)$. For such a reason, the maps $res_{V,U}$ are called *restrictions*.

A *section* of F over U is just an element of $F(U)$, whereas a *global section* is an element of $F(X)$ ("global" since it is a section over the entire space).

5.6.2 From Presheaves to Sheaves

Consider a topological space X, an open set U, an open covering U_i of U and a presheaf F on X (for sake of clarity let us assume that the target category of F is just the category of sets $\mathcal{S}$).

Let now s_i be a family of sections, parametrized by the covering sets U_i. We call this a family of *compatible* sections if, given any two of them, say s_i and s_j, it is the case that they agree on the overlap of their supports – that is, if $s_{i|U_i \cap U_j} = s_{j|U_i \cap U_j}$.

We say that a global section s is the *gluing* of these $s_i^\prime s$ if $s_{|U_i} = s_i$ for any i.

We also say that a presheaf is a *sheaf* if a gluing section s over U exists and is unique for every U. Formally, if $t \in F(U)$ is such that $s_{|U_i} = t_{|U_i}$, then $t = s$.

Notice that F being a sheaf guarantees the existence of a global gluing section for F (a section over the entire space X which glues together sections over open sets of the covering). What this means is that the synthetic operation that glues sections together on the overlaps of open sets extends to the base topology taken in its entirety. Local compatibilities "gather up" progressively all the way to a global synthesis.

5.6.3 Badiou's Synthetic Ontology

Now we are in the position to present a synthesis of Badiou's phenomenology in all its integrity.

Let T be the transcendental seen as a category – with its natural partial order $\leq$ as the only possible morphism between its elements.

Let F be the functor from T to the category of sets S with functions as morphisms. F is defined as follows:

- $F(a) = E(a)$
- Given a morphism $p \to q$, $F(p \to q) = f : F(q) \to F(p)$ with $f(y) = y \mathbin{\int} p$

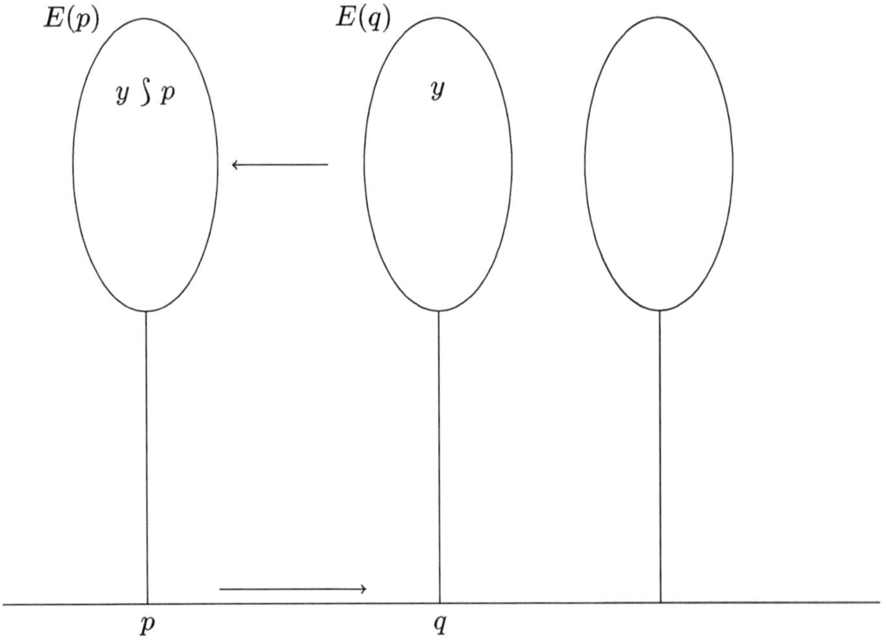

It can be shown that F is a sheaf, and that is Badiou's main claim: localizations play the role of restrictions, sections are just constituents of the world and compatible elements correspond to compatible sections in the language of sheaves outlined at the beginning of this section.

The point might seem to be a trivial one. In the end, Badiou appears to be doing little more than rephrasing a basic construction in category theory within his own philosophical framework. In what follows we discuss the question of how translating this mathematical structure into a philosophical and metaphysical context might be generative of new insights, illustrating both the strength (especially from the perspective of abductive reasoning) and the shortcomings of his program. The key concept that bridges the mathematical formalism and the philosophical interpretation is Badiou's notion of a *point*.

5.7 Points

> A point is a kind of analytic mediation between the transcendental complexity of a world (its often non-classical logic) and the (always classical) imperative of binarity or decision ([6], p. 439).

Badiou's synthetic ontology is really nothing more than a formal framework for posing certain questions and elucidating certain philosophical problems. Nonetheless, Badiou does not leave the formal framework as a mere empty shell. He also introduces his own interpretation of how this framework ought to be filled in with philosophically relevant content. He is especially concerned to treat the sheaf-structure that imposes a coherent system of intra-worldly relationality on the pure sets of extra-worldly ontology as a support for modeling subjectivity.

From the tension between a Boolean (for Badiou often this means binary) and a non-Boolean interpretation of the world, the notion of point emerges as a mediator between these two profoundly different semantics.

In *Logics of Worlds* a *point* is defined as a surjective homomorphism from the transcendental T onto the prototypical Boolean algebra $T_0 = \{0, 1\}$. It is the reverse move that brings us back from the generic onto the crisp, from the semantic of possibilities to that of actual decisions.

Abduction, as conceived from Peirce first and, as we have seen in chapter four, intrinsically embedded into the fabric of *Being and Event*, can then be thought as an oscillatory dynamic between T and T_0, a dynamic that ultimately constitutes the very foundation of any conceivable epistemological model, at least from Badiou's point of view.

In order to gain knowledge according to our abduction model, we need first to open ourselves to the world of the possible, where our crisp intuition may be lost but where truly new knowledge may be gained and formal experimentation can be performed. It seems then for this very reason that the generally non-Boolean structure of topological spaces provides a better class of formal models for abductive reasoning.

On the other hand, Badiou's notion of point can be employed to project down possibilities (i.e., potentially explanatory hypotheses) into a realistic experience and practice, where experiment-checking and actual decisional processes may be put in motion and ultimately resolved. Yet while the basic utility of the distinction between Boolean and non-Boolean (classical and non-classical) relational frameworks for characterizing different aspects of abductive reasoning seems well-grounded, the question remains as to how best to employ this difference so as to most effectively model actual processes of experimental thinking.

Without ever making the mathematical connection fully explicit, Badiou's use of "points" to serve as supports in *Logics of Worlds* for the same kind of subject-models that in *Being and Event* were expressed by Cohen's forcing essentially shifts the same mathematical structure (forcing) from one framework to another (from hierarchies of sets defined on the ordinals to Boolean-valued models). To show this in detail would go beyond the limits of the present work, but the interested reader is encouraged to refer to [7], already discussed briefly in chapter four.

5.8 Transcendental Abduction

The mathematical framework of sheaves that Badiou uses to shape his general phenomenology or, in the Hegelian terms preferred by [8], objective world-logic has in fact an even broader significance than that accorded to it by Badiou himself. The natural emergence of a logic of relations corresponding to Heyting algebras from within the mathematics of sheaves can, and in our view should, be interpreted philosophically as an insight into fundamental metaphysics. And the interplay between non-Boolean and Boolean Heyting algebras in this regard does have direct relevance to the mathematical exploration of the dynamics of being, experience and knowledge. But it would seem better in this regard when advancing into such delicate and untamed territory to err on the side of openness rather than closure, to treat the space thus newly manifest abductively as a space of possibilities ready for exploration rather than a defensive bulwark for arbitrarily imported philosophical views.

One of the core elements of Kant's methodological revolution in philosophy was to treat the intelligible structure of experience not as a simple datum but as a consequence of intrinsic features of the formal conditions of possibility for any conceivable objective experience whatsoever. This form of reasoning constitutes the Kantian "transcendental deduction of the categories".

One of the virtues of the Kantian approach is that the mathematical nature of scientific knowledge becomes on its basis an expected feature of experience (a "matter of course") rather than a Wignerian "miracle" (a "surprise" in need of further explanation). But the mathematics to be found in and as the scientific knowledge of nature is, for Kant, ultimately grounded in classical logic. For Kant, as for the later logicist program in mathematical foundations following in the line of Frege and Russell, mathematics finally is just a systematic explication of fundamental logical principles and their deductively derived consequences. The "transcendental deduction" simply rewrites these formal principles of deductive judgment as necessary conditions of real experience.

An alternate tradition in mathematical foundations represented by, among others, Peirce views mathematics constructively, so that logic does not govern mathematical truth so much as follow upon it as merely one local region. In contemporary treatments, for instance, mathematical logic is developed on the basis of a background set theory, and even though the axioms of set theory itself are typically expressed in standard first-order logic, the actual "work" of proving canonical theorems in mathematical logic (such as Zorn's lemma) depends less on these axiomatic foundations than on intelligently formulated definitions (such as those of *partially ordered set*, *chain* and *upper bound*) that construct well-motivated abstract objects *the relations among which* generate the proof-results sought. The "truth" of such results (in the Hegelian sense of truth as the concrete outcome of a dialectical development internal to some concept or problem) is essentially mathematical and constructive, not logical (in the sense of formal deduction from arbitrary axioms).

In accordance with this latter tradition that as it were prioritizes mathematics with respect to logic rather than the reverse, the method of transcendental deduction may be relatively relaxed to that of "transcendental abduction", in the sense that the application of the forms of logical judgment to the concrete data of experience is no longer itself given as a logically grounded necessity. Instead, the constructive use of mathematics as a purely iconic or structural form of reasoning is conceived as an ineluctable aspect of any epistemic process whatsoever without thereby having to decide in advance what its specific deductive or logical character of reasoning must be. The "logic" of experience becomes an experimental variable without having to abdicate the rigor of mathematical reasoning.

If Kantian method provides a definite philosophical justification and ground for Newtonian a priori science, a less secure but perhaps more realistic method of transcendental abduction promises to motivate and guide procedures of a posteriori experimental science for which mathematical reasoning becomes itself an a priori necessary tool but not a pre-established result. Successful scientific inquiry on this basis is ineluctably constrained to *use* mathematics (in the somewhat broadened Peircean sense of abductive experimentation on ideal iconic models) but this necessary excursus into ideality does not restrict the objects of what science might thereby come to know. The correlates to epistemic states will thus in no way be limited to domains of mathematical forms or ideal laws or anything of that sort. In this respect, then, the methodological framework of transcendental abduction aims to hold open a maximally robust realism, that is, a scientific orientation that need not ever give up on the possibility of genuine knowledge while still remaining de jure unrestricted in its capacity to be "surprised" by the real.

In *Logics of Worlds* Badiou invariably conceives of a given world W as possessing a single relational framework T_W that constitutes the determining "logic" of that world. This logic, as represented by the Heyting algebra induced from the partial order of the world's "degrees of appearing", governs not only the phenomenological relations and overlaps among the diverse objects of W at some fixed time t. Importantly, for Badiou the transcendental T_W regulates and accommodates the representation of the processes of change that occur in W. Yet by maintaining the fixity of the universe of sets as the domain of ontology, change is determined a priori to be nothing more than a variation in appearance or manifestation, never a genuine feature of ontology or reality as such. This is of course entirely consistent with Badiou's commitment to philosophical Platonism. But it is not compatible with either empirical science or the most natural interpretation of a "mathematical ontology" that would be grounded (more generally and less decisionally) in topoi rather than (less generally and more decisionally) in sets.

What Badiou never seems to consider is the possibility of varying the transcendental T_W itself given some fixed world W, or more precisely given the collection of sets representing the underlying ontology of W. Rather, Badiou does in a certain sense consider the possibility of varying T_W, but only in the "downward" or "collapsing" case that maps T_W, however complex its internal structure, into the simplest non-degenerate Boolean algebra $\{0, 1\}$. This collapsing of the possibly quite rich structure of a world's transcendental into the binary simplicity of $\{0, 1\}$ serves as the formal basis for Badiou's doctrine of "points" and ultimately his "materialism". Yet if Badiou's cunning treatment of foundational mathematics as formal ontology is itself treated not as a nexus of philosophical claims to be judged either true or false but instead as a pragmatic tool, a theoretical model (or space of experimental reasoning) that might prove useful or not for working "concretely" with various philosophical questions, then the most interesting possibilities might very well be represented and tested by *extending* and *enriching* the transcendental of a world.

As a philosopher, Badiou is concerned most deeply with the problem of *truth*. This means that from a semiotic point of view, his linguistic statements and his employment of the complex machinery of set theory and other branches of mathematics must be understood as essentially *propositional* in form. As we saw in chapter two, Peirce's insightful analysis of propositional signs (*dicisigns*) as coordinations of iconic and indexical components helps to shed light not just on what propositions are but more importantly how they actually work, their semiotic real/structural conditions of possibility. By fixing a single, invariant transcendental T for each world W, Badiou effectively decides for the indexical aspect of his mathematical ontology and phenomenology to the detriment if not total exclusion of their potentially much more valuable iconic dimension. He thus largely neglects to treat his own formal modeling of ontology and phenomenology *as* a model.

5.9 Conclusion

We conclude that Badiou's use of the mathematics of sheaves to formalize the phenomenal "logic of appearing" of world-bound or relationally embedded entities in fact outstrips his stated purposes and the limits in which he presents his findings. Despite Badiou's own intentions to delimit the role of non-classical and purely relational determinations solely to the phenomenal realm and to preserve the space of ontology as wholly governed by the classical logic of standard set theory, the objective mathematical features of sheaves render the fixed and independent status of a set-theoretical ontology deeply problematic. Once the formal role of sheaves is introduced, there is simply no good reason to decide for sets as opposed to other equally workable topoi.

A conclusion very similar in spirit to ours has been reached independently by Veilahti ([1], p. 4, grammar corrected) in one of the few critical analyses of *Logics of Worlds* that treats Badiou's mathematical tools in detail.

> Lacking categorical insight, however, neither Cohen nor Badiou could figure out how to think of such contradicting models or situations together. Badiou's philosophy could only deal with the occurrence of the "inconsistent" as a "generic" decision, that is, as a choice of only one

> among all situations [...]. He thus believes that one needs to choose which context to inhabit instead of residing in and between many of them all at once. Topos theory, by contrast, does use categorical techniques to specifically express the amalgam of such situations so that the need to decide does not arise but possibly afterwards. Badiou's decision between situations then only emerges as a (local) projection of that topos onto set theory. Such projections [...] are in fact specific kinds of geometric morphisms that topos theorists refer to as "points". It is not by accident that a "decision" in Badiou's vocabulary then stands for that precise procedure through which category theorists "make a point". There is much synergy between dialectical and scientific modes of materialisms.

In the above work [1] the author presents a very thorough analysis of the technical aspects of *Logics of Worlds*, and he shows that Badiou's attempt to position the logic of appearing within a purely relational framework stops short from being complete. In particular, it is emphasized that Badiou ultimately works in "local" topoi, and that his limited use of the idea of functors, as well as that of the Grothendieck sheaf, is structurally flawed.

In spite of his shortcomings, however, Badiou has succeeded in opening up a profound channel of communication between modern mathematics and philosophy, with largely unrecognized potential.

Badiou's overall project linking *Being and Event* and *Logics of Worlds* consists of the translation of certain sophisticated mathematical techniques into the philosophical registers of ontology and phenomenology. In particular, as analyzed in chapter four, Cohen's set-theoretical technique of forcing becomes under Badiou's ontological interpretation a means for modeling the creative aspect of abductive reasoning. In that context, epistemological and ontological concerns are no longer merely coordinated but in fact coincide due to their representation in and as set theoretical formalism. In the present chapter, we have seen how Badiou supplements the set-theoretical ontology of *Being and Event* with an essentially sheaf-theoretical phenomenology in *Logics of Worlds*. Here, the representation of coordinated systems of local-global relations as supported by the mathematics of sheaves suffices to model the relational logic of "appearance" or "manifestation" that characterizes phenomenology.

What is most interesting and provocative in this two-pronged mapping of mathematics into philosophy is the interplay between the two levels of ontology and phenomenology themselves. If the decision to interpret set theory as ontology and sheaf theory as phenomenology is understood as a sort of working hypothesis, Badiou's project then appears at its most general level as an experiment in model-based reasoning that takes the most sophisticated mathematical environments available to us at the current intellectual conjuncture as optimal testing-grounds for the philosophical investigation of fundamental questions of being, appearance and truth.

The shape of reasoning expressed in Badiou's project is in this way thoroughly iconic in character. Yet the iconic and abductive reasoning that underlies Badiou's overall philosophical project is, ironically enough, subordinated to an a priori philosophical decision for Platonic metaphysics that Badiou makes in advance of his actual mathematical investigations, at least his explorations of sheaves and topoi. In this way, Gironi's critique of Badiou's *Being and Event* from a realist and naturalist perspective as discussed at the close of chapter four is valid a fortiori for *Logics of Worlds* as analyzed here. The argument developed in the present chapter has been mounted from a point of view that might be deemed *mathematical naturalism* in the sense that it reasons from a standpoint that would take the mathematical modeling of conceptual and philosophical questions to be genuinely abductive and diagrammatic in character, namely as capable not only of representing what is assumed philosophically or conceptually in advance but of discovering newly relevant information about the theoretical object at stake and possibly subjecting the initial assumptions in play to critical revision based on objective features of whatever mathematical domains are employed as diagrammatic models.

References

1. A. Veilahti, Alain Badiou's mistake: two postulates of dialectic materialism. arXiv:1301.1203v3
2. A. Badiou, *Logics of Worlds* (Continuum, London, 2009)
3. G. Brady, T.H. Trimble, A categorical interpretation of C.S. Peirce's propositional logic Alpha. J. Pure Appl. Algebr. **49**, 213–239 (2000)
4. J. Woods, Ignorance and semantic tableaux: Aliseda on abduction. Theoria **60**, 305–318 (2007)
5. F. Stjernfelt, *Diagrammatology: An Investigation on the Borderlines of Phenomenology, Ontology, and Semiotics* (Springer, Berlin, 2007)
6. A. Badiou, *Being and Event* (Continuum, London, 2005)
7. J. Bell, *Toposes and Local Set Theories: An Introduction* (Dover, New York, 2008)
8. A. Rodin, *Axiomatic Method and Category Theory* (Springer, Berlin, 2014)

Chapter 6
Abductive Realism in Topos Theory

6.1 Introduction

The foundation of Badiou's ontological schema relies on the notion of a transcendental T. As emerged from the analysis in the last chapter, T is in fact best understood as a two-fold entity: It is used to measure the degree of identity of relations between objects in the world, and, on the other hand, is a structured system of relations by itself, namely a complete distributive lattice (Heyting algebra), with the property of being essentially optimal with respect to the contexts it is meant to model. The problem that, we believe, the logic of abduction can clarify with respect to this framework is that of reconciling a broadly "realist" perspective with an ontological model that not only admits multi-valued semantics beyond $\{0, 1\}$ but, more importantly, revolutionizes the classical logic of truths by including non-Boolean constructs, as for instance, admitting situations where the law of the excluded middle is not generally valid.

The problem of *realism* in philosophy of science is a deep and relatively long-standing one that has received attention from a variety of quarters. Quantum mechanics remains a canonical test case for realist conceptions of science because it (1) enjoys tremendous predictive success and yet (2) does not easily suggest, to put it mildly, any straightforward coordination of its formal schemes of representation with seemingly uncontroversial notions of what characterizes physical reality. Among the well-known aspects of this latter problem are such features of quantum mechanical mathematics that appear to entail the rejection of determinate characteristics of reality at all times such as the irreducibility of non-local phenomena as expressed by Bell's theorem. Because of (1), quantum mechanics aligns readily with instrumentalist conceptions of science. Because of (2), quantum mechanics provides an obstruction to a straightforward realist mapping between theory and reality.

For our purposes, the most important aspect of the debate is the following: Realism requires at a minimum that relevant aspects of scientific theories (typically, but not necessarily, *terms*) refer in some way to relevant aspects of reality (typically

G. Caterina and R. Gangle, *Iconicity and Abduction*,
Studies in Applied Philosophy, Epistemology and Rational Ethics 29,
DOI 10.1007/978-3-319-44245-7_6

individuals or *objects*). The reference relation in question is usually taken to be one that involves mappings of distinct terms in the theory to distinct real entities (whether directly observable or not), but many other relations are possible, including those proposed by various forms of structural realism. In any case, what any purportedly realist interpretation of quantum mechanics must itself come to terms with is an apparent disconnect between the formal structures inherent to the scientifically effective mathematics and any intuitively reasonable conception of reality's structure (ontology) and furniture (ontics). In our view, this problem is significantly clarified when considered from the standpoint of the relation between abduction and iconicity that has occupied us up to this point.

The problem of realism with respect to quantum mechanics serves as an especially vivid instance of the more general problem of how and why mathematical structure serves as a uniquely effective framework for scientific abduction. Therefore the question at the heart of this last chapter will be an analysis of the concept of realism within abductive processes, carried via the study of three recent attempts (Flori, Epperson–Zafiris, and Ellerman) to restore realism in the formalism and interpretation of quantum mechanics by way of categorical methods.

Although these three approaches differ in spirit and methodologies (Flori and Epperson–Zafiris make use of topoi, while Ellerman bases his thesis on the partition logic he invented), all of them share that the use of certain mathematical notations enables a realistic perspective intended to resolve some of the apparent discrepancies between the different interpretations of quantum mechanics. For a helpful survey of and introduction to the main types of mathematical formalism used to represent quantum phenomena (one which, however, neglects to discuss the approach via sheaves and topoi) see [1]. Also relevant to this discussion is [2], which raises a variety of more technical issues and provides a nice survey of the main open lines of research involved in linking the categorical approach to quantum mechanics to issues in abstract diagrammatic reasoning and compositional semantics, topics on which we will largely remain silent here for reasons of economy.

To demonstrate the more general point about categorical mathematics, abduction and scientific realism, the following sections introduce several formal constructions. The section on Flori explains one of the most important results in category theory, Yoneda's lemma, which we introduce as a generalization of Cayley's theorem. We use this result to characterize how the subobject classifier is generated in a presheaf topos from sieves, which Flori uses to generate "realist" truth-values for physical quantities that respect such "paradoxes" as non-locality. In the section on the relational realism of Epperson and Zafiris we discuss the sheaf-theoretical notion of germs over a point in a topological structure. Here too a realist interpretation of quantum mechanics is affirmed on the basis of the topos-theoretical mathematics. And finally in the context of the discussion of Ellerman's approach we summarize his formulation of partition logic which, while not strictly speaking categorical, is certainly inspired by the relational and algebraic methods endemic to category theory.

Our overall claim in this chapter is that the internally relational standpoint that characterizes category theoretical mathematics lends itself readily to the problem of realism in quantum mechanics precisely because the role of mathematics in this context is abductive in character in a way that is itself naturally expressed categorically. When the use of mathematics in natural science is aligned neatly with our ordinary ontological and epistemological prejudices, its abductive role tends to be far less salient. It is only when the abductive movement instantiated in the relation between the epistemological and ontological sides of a scientific theory as mediated by its mathematical formalism in fact challenges ordinary assumptions that the key role of mathematics in establishing (and not merely representing) a potentially realist scientific orientation becomes especially important. Interestingly, this perspective entails that ontological and epistemological concerns in science, while distinct, are nonetheless correlative to one another and, more deeply still, are both equally grounded in the formal mathematics that in a certain sense generates them as necessary polarities of any scientific realism.

6.2 Topos Quantum Theory: Flori

The first of our three "models" of realism in quantum mechanics based in categorical mathematics is that of Flori in [3], which draws on earlier work from Isham [4]. We focus here on the role of Yoneda's lemma in her detailed and comprehensive presentation. Yoneda's lemma provides us with a synthetic representation, within the category of sets, of the functorial structure of any category. In this section, we will show how this result is crucial to the goal of producing a realist interpretation of quantum mechanics (in the sense indicated above) in Flori's approach.

Flori's overall strategy is to take the mathematical framework that, in her view, determines a classical theory as realist and, while retaining the properties that define its realist kernel, to resituate these in a more general and internally variable topos-theoretical environment that is as such more amenable to quantum mechanics. According to Flori [3], p. 10, the intrinsically realist orientation of classical theory with respect to a given system depends upon the following theoretical elements:

1. A clearly defined *state space* S such that at any specified time t the system's properties are exhaustively described by a unique element of S.
2. Determination of physical quantities by real-valued functions.
3. Propositions about the system understood as definite subsets of the state space S taking values within some specified interval for any given physical quantity.
4. A Boolean and 2-valued logic governing such propositions.
5. Probabilities as expressive only of epistemic indeterminacies, not real ones.

In different ways, each of these classical realist elements faces an intrinsic challenge in quantum mechanics as usually formulated. As Flori points out, the canonical mathematical formalism for quantum theory leads naturally to a strictly instrumentalist and therefore anti-realist interpretation. This is due particularly to the internal division *within the formalism itself* between measuring apparatuses and systems subject to measurement ([3], p. 10):

> In fact the formalism of quantum theory implies a clear distinction between measuring apparatus and measuring system, such that the act of measuring gets ascribed a special status. In this setting, measurement becomes a means for assigning a probabilistic spread of outcomes rather than a means to determine properties of the system. Thus the very concept of properties ceases to have its common sense meaning, since its definition is now intertwined with the act of measurement. It is as if properties acquire the status of latent attributes which are brought into existence by the act of measurement, but which can not be said to exist independently of such measurement. Therefore it becomes meaningless to talk about a physical system as possessing properties.

Flori's point here is that the internal structure of the standard formalism requires a clear difference between two types of represented aspects of the system: measurement apparatus and bearer of measurement. If this formalism within the theory is taken in a realist fashion as representing the actual aspects of the real physical system at issue, then at least one of the core notions of realism (on her view) dissolves. Thus the formalism itself appears to set up a contradiction for any viable realist interpretation: if the theory is interpreted in a realist manner, then the reality thereby represented by it cannot be made objectively intelligible.

Part of Flori's strategy to recover a realist interpretation of quantum mechanics involves recasting the very notion of mathematical representation in categorical and sheaf-theoretical terms via the internal logic of topoi. The following sections trace this fundamental shift from "external" to "internal" mathematical representation.

6.2.1 Naturalizing Abstraction: Cayley's Theorem

As already discussed in our first chapter, the problem of representation lies at the core of mathematical thinking and practice. Here the "problem of representation" signifies the search for formal domains that are structurally rich enough to translate, and possibly simplify, the nature of a given mathematical construction.

A basic but fundamental example of representation is given by Cayley's theorem in the context of group theory. It shows that every group can be realized as a subgroup of a large enough group of permutations. In what follows we will discuss both why this result is relevant and what its role is within a purely relational context.

Groups are just sets, which we can denote by G, whose elements are related by a binary operation $\star : G \times G \longrightarrow G$ such that $\star$ is associative, an identity element with respect to $\star$ exists (let us denote such element by e) and such that an inverse element exists for every g (let us call the inverse g^{-1}). It is easy to prove that g^{-1} is both a right and a left inverse for g and that e is unique.

Familiar examples of groups are the integers with the operation of addition, the set of $n \times n$ matrices with the operation of matrix multiplication, the set of the multiples of p, where p is a prime number, with the operation of multiplication, and the set S_n of all bijections of a set with n elements taken together with the operation of composition of functions.

This latter group S_n (without loss of generality we can assume n finite) plays a very special role in the problem of representability. Indeed, Cayley's theorem shows that any group whatsoever can be mapped isomorphically into some S_n. Why is this interesting?

First of all, subgroups of S_n emerge very naturally when looking at concrete symmetries of geometrical objects in space. More importantly, bijections of a group into itself emerge canonically as a concrete representation of the group in itself, according to what we might call "the canonical double role of group elements".

Consider a group $(G, \star)$ and fix an element g. Such a g induces a map $\hat{g} : G \longrightarrow G$ which takes an $x \in G$ into $g \star x$, that is $\hat{g}(x) = g \star x$. Therefore g can be understood as playing the double role of both an element constitutive of the group, and also an operational role as an endofunction from G into G.

It turns out that, since G is a group, $\hat{g}$ is a bijection! Two more good things happen. The first is obvious: the set of all $\hat{g}$s, as we vary g within G, forms a group. Moreover, it can be shown that $\hat{}$, as a map $\hat{} : G \longrightarrow Sym_G$, is a group homomorphism.

This essentially concludes the proof of Cayley's theorem and, most importantly, gives us some insight into the way certain mathematical structures are endowed with an intrinsic way to both represent and be represented in some canonical context.

6.2.2 *Yoneda's Lemma and Embedding*

Let us move from groups towards a much more general context, that of categories. Given a category $\mathcal{C}$, we can, as we did with groups, exploit the ability of the objects of $\mathcal{C}$ to act on $\mathcal{C}$. Pick an object, say A and associate to any $X \in \mathcal{C}$ the set of arrows going from A into X. This defines a functor $H_A : \mathcal{C} \longrightarrow \mathcal{S}et$, that is, define an object H_A in the category $\mathcal{D}$ of all the functors $\mathcal{C} \longrightarrow \mathcal{S}et$ where arrows are assumed to be natural transformations.

Yoneda's lemma stems from a result showing that, given any functor $F : \mathcal{C} \longrightarrow \mathcal{S}et$, that is any object F in $\mathcal{D}$, the set of all the arrows in this category between H_A and F, that is, $H_{H_A}(F)$, is isomorphic to $F(A)$. Notice that this latter isomorphism lives in $\mathcal{S}et$.

As a corollary of this result, we obtain Yoneda's lemma, which can be understood as a more general, relations-only version of Cayley's theorem: the functor $H : \mathcal{C} \longrightarrow \mathcal{D}$ is full and faithful, which really means that it is a (sort of) bijection between $\mathcal{C}$ and $\mathcal{D}$. Bingo! We found that any category $\mathcal{C}$ can be thought/seen as a subcategory of functors from $\mathcal{C}$ into $\mathcal{S}et$. This completes the analogy with Cayley's theorem: we have a canonical "larger" category ($\mathcal{D}$), which retains some basic information on $\mathcal{C}$ (it is indeed the category of functors $\mathcal{C} \longrightarrow \mathcal{S}et$) in which $\mathcal{C}$ is "exactly" represented. For further details, see Appendix A.

6.2.3 Sieves

From the problem of representation we pass to the problem of propositional evaluation, the categorical analogue of which is the role played by sub-object classifiers in topoi. The existence of a sub-object classifier is a fundamental structural feature of any topos. When we talk about the *internal logic* of a topos, we really refer to the fact that, as a category, the structure of a topos is rich enough for a sufficiently robust logic to emerge from it. The semantics of such logic is reflected into the elements of the sub-object classifier—in simple words, the elements of the sub-object classifier can be regarded as the distinct truth values of the topos logic.

It is therefore only natural to look at the structure of the elements of sub-object classifiers, and in order to do that we will need a finer analysis of certain topos-theoretical objects.

Definition 7 A **sieve** on an object $A \in C$ is a collection S of morphisms in C whose codomain is A and such that, if $f : B \longrightarrow A \in S$ then, given any morphisms $g : C \longrightarrow B$ we have $f \circ g \in S$, i.e. S is closed under left composition:

Figure 1

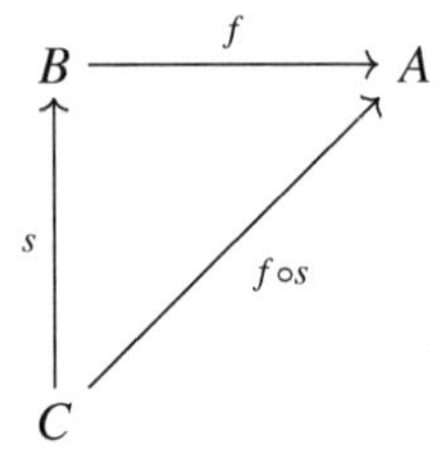

Behind this somewhat technical definition there is a simple intuition: a sieve on a object of C is just a "compositionally closed" collection of morphisms that "absorbs" subsets of morphisms into that object, under composition. A sieve is therefore the analogue of *ideals* for rings and of *filters* for topological spaces.

If we denote by $\Omega(A)$ the collection of all the sieves on an object A, any arrow $f : A \longrightarrow B$ induces an arrow $\Omega_{BA} : \Omega(B) \longrightarrow A$ in a natural way:

$$\Omega_{BA}(S) = \{g | cod(g) = A \ and \ f \circ g \in S\}$$

where $S \in \Omega(B)$. Among all the sieves on A, there is the principal sieve, denoted by $\downarrow A$, which is the sieve that contains the identity arrow on A. Clearly $\downarrow A$ coincides with the set of all the arrows into A.

A crucial property of $\Omega(A)$ is that it can be endowed with a Heyting algebra structure, and the key to understanding why this is the case is given by the Yoneda lemma.

6.2.4 The Presheaf in the Topos of Quantum Mechanics

Theorem 7 *Given an object $A \in \mathcal{C}$, (where $\mathcal{C}$ is a locally small category), then a sieve on A can be identified with the representable functor (as defined in Appendix A) $y(C) = Hom_{\mathcal{C}}(-, A) \in Sets^{\mathcal{C}^{op}}$.*

For the proof of this theorem, see [3], p. 88.

Summarizing the situation so far, we have seen that we can attach, to any object A of any (small) category $\mathcal{C}$ the set $\Omega(A)$ of all the sieves on A, and that such a set $\Omega(A)$ forms a Heyting algebra.

$$A \longrightarrow \Omega(A)$$

Given the natural contravariance intrinsic in Ω, we can ask the question of whether Ω is an actual functor. The answer is yes, and therefore Ω can be regarded as a presheaf from $\mathcal{C}$ to the category of sets.

Let us focus our attention now on the category whose objects are all the presheaves on $\mathcal{C}$, that is all the contravariant functors from $\mathcal{C}$ into **Sets**, in which the arrows correspond to natural morphisms between presheaves, and let us denote this category by $\mathbf{Sets}^{\mathcal{C}^{op}}$.

Theorem 8 *Ω is the subobject classifier of* $\mathbf{Sets}^{\mathcal{C}^{op}}$.

Notice that Ω is a local-to-global kind object. The arrows correspondent to true, for instance, will be given by $\top : 1 \longrightarrow \Omega$, which is the natural transformation that has components $\top_A : \{\star\} \longrightarrow \Omega(A)$ given by $\top_A(\star) = \downarrow A$, where $\downarrow A$ denotes the principal sieve on A.

Technical details aside, the fact that the category of presheaves into **Sets** is a topos, is central to Flori's development of a comprehensive categorical language of Quantum Mechanics. An appropriate choice for the category $\mathcal{C}$ in $\mathbf{Sets}^{\mathcal{C}^{op}}$ (which she shows to be the category of abelian algebras), enables each quantum object to be defined locally. The interesting point is that locally, in this precise sense, quantum mechanics is shown to behave classically even though globally it does not. Moreover, the collection of these local, classical snapshots is proven to be ordered by inclusion, implying the fact that "although one defines each quantum object locally, the global information is never lost, since it is put back into the picture by the categorical structure of the collection of all these classical snapshots. Hence the task is to find a topos which allows you to define a quantum object as (roughly speaking) a collection of classical approximations" ([3], p. x).

6.2.5 Truth Values in a Topos

One of the main goals in Flori's work is the construction of a logical apparatus whose semantic is—at least from her perspective—a "best fit" for evaluating quantum propositions. This logical apparatus is shown to be a topos, a category which has a rich enough internal structure to generate its own logic and, therefore, its own algebra of truth-values, that is the algebra of the elements of the subobject classifier.

This is a first important point from an iconic-abductive perspective: truth values are generated from within a relational context (so they are not given as a priori algebraic structures such as multivalued algebras in Lukasiewicz logic). Moreover, they are in general not only multi-valued but also non-Boolean: in a general topos, the algebra of elements of the subobject classifier is a Heyting algebra, and therefore classical constructs such as that of the excluded middle may cease to hold.

Ultimately, Flori points out that "The important feature of these new truth values is that, given any set of quantum propositions, even incompatible ones, it is always possible to assess their truthfulness simultaneously", and that "Moreover the set of truth values forms a Heyting algebra thus, we obtain an intuitionistic logic. What this implies is that a more realist picture of quantum theory emerges" (p. 177).

While these claims, especially to the extent they seem to suggest a consequential relation between truth-values being a Heyting algebra and topos QM being realistic, can be understood as subjective and in some sense arbitrary, what makes her work relevant to us is the notion that, when lifted into a topos-theoretical framework, quantum mechanical entities and processes (including, for instance the notion of measurement apparatus), along with their syntactical and semantic representation, share the same ontological status (as object/relations in the topos).

6.3 Topos Quantum Theory: Epperson and Zafiris

An approach to quantum theory via sheaves and topoi very much akin to Flori's in its formal dimension has been developed by Epperson and Zafiris in [5]. Much as for Flori, their motivation for shifting mathematical terrain to topos theory and the internal Heyting logic of sheaves as a way to represent quantum theory stems from a desire to preserve a *realist* interpretation of quantum mechanics. Yet interestingly, their understanding of what exactly "realism" means in this formal context is quite different. Instead of a conception of realism grounded in a scientific epistemology of quantifiable measurement, Epperson and Zafiris advocate a metaphysics of what they call "relational realism", that is, a philosophical realism emphasizing its ontological dimension. The stress is placed on "relational" in the name relational realism. What is most crucial to their position is the claim that reality itself is relational in essence. The proper domain of ontology is not merely the class of existent objects (a notion that would be captured well by the mathematics of sets, as in Badiou) but rather the relations that ultimately link aspects (parts) of reality to one another. Determinate epistemic considerations then flow naturally from this relational conception of entities.

6.3.1 Ontological Duality

Important for their approach is the distinction between two types of consequence-like relation, "the asymmetrical orders of syntactic, Boolean material implication, and semantic, logical implication (i.e., entailment)" ([5], p. 140). The intuition behind this distinction is not a difficult one, but the unfolding of its relevance both within the mathematical formalisms of quantum theory and for the interpretations of those formalisms is quite surprising. Roughly, the difference between what Epperson and Zafiris call the "syntactic" and "semantic" modes of implication corresponds to the difference between the "form" and the "content" of theoretical propositions. Syntactic, or material implication concerns only the external or purely formal relations between the truth-values of propositions joined syntactically by a non-commutative binary operator. The possible range of these relations amounts to the standard truth-tables for Boolean logic. The operation is purely combinatoric, ranging over possible truth-values for ordered pairs of propositions and producing as output the same values for the same inputs regardless of the specific features of whatever semantic domain is in play in any particular propositional context.

The semantic, or what Epperson and Zafiris call (somewhat confusingly) "logical" implication concerns, on the other hand, the specific content of the propositions that are joined by a different type of binary operator, in fact, really a relation that either holds or does not hold between any two given propositions. This operator measures the extent to which a given proposition, if true, would "force" some other proposition to hold. "Socrates is a man" would imply, in this semantic sense, "Socrates is mortal".

As grounded in this elementary distinction, the viewpoint of Epperson and Zafiris's relational realism is quite close to that of the various decoherence-based interpretations of quantum physics, while diverging from such decoherence-based approaches at certain key junctures. One element that the different approaches share is a relatively "ignorance-preserving" position with respect to the factual "data" of quantum events. As they put it ([5], p. 168–9),

> neither the decoherence-based interpretations nor the relational realist cosmology proposes a mechanism that "accounts for" the existence of quantum events, the "objective facts" of reality. Unlike other ontological interpretations of quantum mechanics that consider this a deficiency or a problem to be solved, the decoherence-based interpretations, whether implicitly or explicitly, acknowledge that facts are necessarily presupposed by quantum mechanics and indeed [...] by the scientific method in general. Focus is instead placed upon the *forms of relations* among quantum events - the logically conditioned evolution of potential relations to probable relations - and the conception of novel quantum events as predicative facts generated by these relations.

The connection here to the theme of abduction should be evident, and the reference to scientific method is not an arbitrary one. Epperson and Zafiris recognize *within* the reconstructed quantum formalism they propose an abductive or quasi-abductive structure. Just as scientific abduction differs from deduction by, among other things, taking a given *fact or collection of facts* (a surprising event) as "input" rather than a collection of already syntactically formatted premises, the interpretation of quantum theory they propose treats concrete physical facts not as formal elements of the theory but rather as irreducibly "real" data *for* the relations expressed by the theory.

What emerges from this interpretation of quantum mechanics is an ontology structured through an irreducible duality. In particular, this duality appears as the impossibility of deciding for either the discrete or the continuous as constituting the ultimate foundational level of reality.

Epperson and Zafiris express this point by claiming that ([5], p. 169)

> although the quantum actual occasions themselves are discrete, their *potential relations* are continuous and therefore metrically describable in terms of their extensive features. Thus, in the same Aristotelian sense that potentia always presuppose actualities, the continuous metrical descriptions of extension deriving from these potential relations presuppose a discrete mereotopological extensive order of actual facts. But at the same time, in the relational realist philosophical scheme it is also true that actualities always presuppose potentia, since potential relations are always generative of novel actualities [...]. In this regard, the various attempts by the physical sciences to depict physical reality as either fundamentally discrete or fundamentally continuous (reflected, for example, by the tension between quantum mechanics and general relativity) neglect the fact that quantum actual occasions cannot be abstracted from, or even defined without reference to, their actual *and* potential relations, both logical/mereotopological and extensive/metrical.

This essential duality, which is not strictly speaking a dualism since the terms are mutually implicated and thus cannot be defined or conceived independently of one another, constitutes the most striking feature of the realism they espouse. Philosophically, this metaphysics and ontology is derived from Whitehead, especially [6].

The ontological upshot from this point of view is that objects and relations must be understood as thoroughly interpenetrating one another without either of these semantic terms being wholly reducible to the other. Thus, despite their defense of a "relational realism" Epperson and Zafiris are not somehow claiming that "only relations are real". Rather, their claim is that relations are in a rigorous and justifiable sense *just as* real as actualized entities. Much as the concepts of North and South poles of a magnet are meaningless when conceived as fully independent of one another and yet are by no means thereby identified, the Whiteheadian ontology underscored by their interpretation of quantum mechanics asserts an irreducible duality (not a dualism) of actuality and potentiality, of concretized, relation-supporting objects and concretizable, object-dependent relations ([5], p. 169):

> Potentiality and actuality are thus mutually implicative in the relational realist philosophy, whose fundamental objects are quantum mechanical units of logico-physical relation rather than simply units of physical relata. Objects are therefore always understood as relata, and likewise relations are always understood objectively.

Rather than trying to evaluate the provocative ontological picture given here as adequate or not from a philosophical (or, for that matter, scientific) perspective, from the present standpoint what is most important is the way that this relational ontology is at once grounded in and derived from the recasting of quantum physics in a topos-theoretical framework. In other words, what is salient here is the way that the overall claim of Epperson and Zafiris regarding the consistency of a Whiteheadian ontological interpretation of quantum theory is itself produced abductively on the basis of the internally relational or "arrows-only" orientation of category theory and topoi. The mathematics employed here is not only formally representational, but suggestively generative of ontological interpretations that play the role, from this perspective, of more or less likely hypotheses. It is thus not simply a matter of producing arbitrary theories or interpretations as constrained merely to be compatible with observational and theoretical data, but of using the iconic characteristics of such data to determine a space of motivated "hypotheses" that conform iconically to what must be explained or understood. One relatively straightforward example is provided by the central role accorded in the category theoretical formalism used by Epperson and Zafiris to *germs* over a generalized topology. We sketch this construction in the following section.

6.3.2 Points, Fibers, Germs

The crucial shift from the standard formalism for quantum mechanics to the relational realist one advocated by Epperson and Zafiris involves, just as with Flori, treating all the mathematical structures relevant to the representation of physical reality as varying via controlled fibers over the ordered elements (the open sets) of a topology. The interplay between the inclusion ordering among these elements on the one hand, which passes from local to global structure, and the controlled restriction ordering among the fibers over them, which moves from global wholes to local restrictions, is of course the now-familiar duality of epistemic/ontic relations that we have examined already in a variety of contexts.

For Epperson and Zafiris, this reconstruction leads naturally to a shift in interpretation from an object-based to an event-based ontology.

To show this, we follow three steps:

1. A formal definition of germ over a topological space.
2. The categorical generalization of the germ notion.
3. The ontological interpretation of a germ-based rather than point-based evaluation.

Definition 8 Given a topological space X, a point $x \in X$, a set Y and two functions $f, g : X \longrightarrow Y$, we declare f and g to be equivalent, written as $f \sim_x g$ if there exists a neighborhood U of x such that f and g agree on U, that is such that $f(y) = g(y)$ for all $y \in U$. The relation $\sim_x$ is an equivalence relation on the set of all functions defined on X and the *germs* at x are defined as the equivalence classes under $\sim_x$.

The main ingredient behind the notion of germ is *locality*: two functions f and g are essentially the same (from the point of view of x) if there is a neighborhood of f where they agree. The shift in the perspective should be clear: we move from a point-wise to a neighborhood-wise concept of equivalence in the fibers (fibers, in this case, are just functions on open sets).

More generally, the notion of germ can be defined in a much more abstract way, relative to any set-valued presheaf defined over a topological space.

Definition 9 Given a topological space X, let us denote by $\mathcal{O}(X)$ the category of its open subsets. Consider a presheaf F

$$F : \mathcal{O}(X) :\longrightarrow \textbf{Sets},$$

a point $x \in X$ and two neighborhood U, V of x. Given two sections $s \in F(U)$ and $t \in F(V)$, we declare s and t to be equivalent, written as $s \sim_x t$ if there exists an open neighborhood of x, say W, such that $W \subseteq U \cap W$ and such that s and t agree on W, that is $s_{|W} = t_{|W}$. The relation $\sim_x$ is an equivalence relation on the set of all sections defined at x and the germs at x are defined as the equivalence classes under $\sim_x$.

It is important to note that the base topology over which the fibers of stalks of germs vary should *not* be identified with continuous regions of space or space-time. In fact, the formal sophistication of the approach used by Epperson and Zafiris depends upon the unique power of sheaf theory to elaborate layers of interdependent relational structures that exist in relative abstraction from any particular "material" data such as sets or functions. Thus, ultimately the germs that are directly employed by the fully reconstructed theory are not derived from continuous real-valued functions but are rather localizations of variable Boolean algebras that vary over Grothendieck sites (generalized topological spaces). Regardless of this additional layer of complexity the details of which exceed the presentation here, the key difference between a germ-based and a point-based formalization of quantum theory from the standpoint of their respective ontological interpretations should be clear. For a point-based evaluation, the question of what is true at a given point is certainly a meaningful question mathematically and would seem to translate to a meaningful question ontologically or "realistically". Regarded, however, via a germ-based formalism the question of what is true at a point is simply inadmissible as a relevant question. A function certainly takes a value at a given point, but what "occurs" at a germ of functions over that point is a different matter entirely. The germ-based interpretation is intrinsically contextual. What is true at a point is determined by what occurs (or can occur) locally in the neighborhood of that point. Such truth cannot be defined in isolation from the ambient mereotopological "space" of other potential contextually-related truths. Thus, the natural connections between distinct contexts are modeled by topological "gluing" operations rather than discrete, combinatorial logics. The formal mathematics thus leads naturally to certain definite ontological conceptions of how reality itself must be conceived to operate at its most fundamental level of objects and relations.

6.4 Quantum Theory and Partition Logic: Ellerman

Flori on the one hand and Epperson and Zafiris on the other preserve, albeit in different ways, realist interpretations of quantum physics by lifting the requisite formalism into a more highly structured space, namely the mathematical framework of topoi. In this sense, their respective topos theoretical reformulations of quantum theory may be understood to preserve the relevant iconicity of the mathematical structures pertaining to quantum physics while allowing the enriched structure of topoi to supervene upon these so as to resolve, from the standpoint of their physical and ontological interpretation, mathematical representations that would otherwise appear impossible or "paradoxical".

A different and in a certain sense "dual" strategy for conceiving a realist ontology that would be compatible with quantum physics is found in the work of David Ellerman. This less technical but still in its own right rigorous project aims to solve the problem of how to reconcile the well-known "paradoxes" generated by quantum theory as applied to physical phenomena with the background ontological/epistemological framework that we use as cognitive agents to grasp and make sense of experience in general. Rather than extending the mathematical framework of quantum mechanics into the richer space of topoi, thereby preserving realism by modifying the underlying logical framework presumed to govern the real, Ellerman takes the intuitive Boolean and set-theoretical mathematical framework for worldly logic as given and instead demonstrates how the paradoxes that result from the dominant interpretations of quantum mechanics may be finessed by treating them as more sophisticated versions of intuitively accessible problems sharing a common structure that can themselves be resolved through straightforward mathematical reformulations. What varies in this case is not the logical infrastructure of the ontology, but rather the mathematically-formulated content, the "grain" of the structurally-determined inhabitants of that infrastructure. Consider the following diagram, where the arrows informally represent projections from a more finely differentiated mathematical environment into a coarser one.

$$\begin{array}{c} Topos\ QM\ (sophisticated\ realist) \\ \downarrow \\ anti-realist\ QM\ with\ paradoxes \\ \downarrow \\ Ellerman\ (simple\ realist) \end{array}$$

The main point is that while these two strategies for handling the "surprising event" of quantum mechanics move as it were in opposite directions, they may both equally be understood as ways of understanding and explaining that work by preserving iconicity in the relevant sense.

In the following section we present a streamlined version of Ellerman's Logic of Partition, which is used to illustrate his more general point.

6.4.1 Partition Logic

Ellerman's original insight is the observation that the Boolean logic of subsets of a given universe set U can be used, essentially, as a model for classical logic. This is a non-controversial claim, as remarked, for instance, in the following quote by Mac Lane and Moerdijk ([7], p. 48):

> The propositional calculus considers Propositions $p, q, r, \ldots$ combined under the operations "and," "or," "implies," and "not," often written as $p \wedge q$, $p \vee q$, $p \implies q$, and $\neg p$. Alternatively, if $P, Q, R, \ldots$ are subsets of some fixed set U with elements u, each proposition p may be replaced by the proposition $u \in P$ for some subset $P \subset U$; the propositional connectives above then become operations on subsets: intersection $\cap$, union $\cup$, implication ($P \implies Q$ *is* $\neg P \cup Q$), and complement of subsets.

The modern treatment of logic can be seen as a generalization of this basic insight. William Lawvere and others have shown how the primitive notion of subset can be generalized to that of *subobjects*, opening the way to the field of *categorial logic*, where a framework within the logic of subsets is replaced by that arising from relations among subobjects. This, in turn, gives rise to the notion of a *topos*—a category rich enough to generate its own internal logic (which reduces to the classical logic of subset inclusion in the special case where we work within the category **Sets**).

On the other hand, given the lattice of subsets of a given universe set U (this is just the set of parts of U ordered by subset inclusion), there is a natural "dual" to it, which is the set of partitions of U.

Before giving the formal definitions, the following example should clarify the idea. Consider first the set $U = \{a, b, c\}$ along with its induced lattice of subsets:

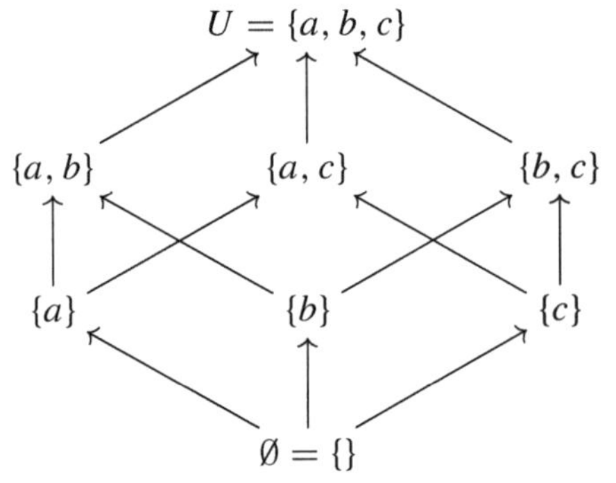

Consider now the dual lattice whose elements are all the possible partitions of U:

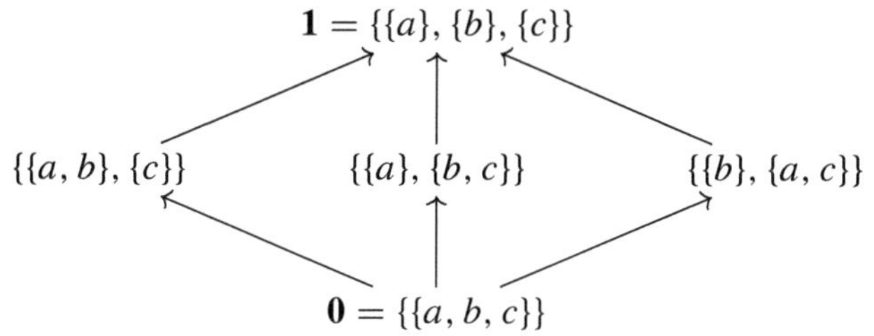

Formally, a *partition* π of U is simply a subdivision of U in equivalence classes, that is in a family of subsets $\{U_i\}_{i \in I}$ such that for any two indices $h, k \in I$ we have that $U_h \bigcap U_k = \emptyset$ and the union of all U_i is equal to U.

Given two partitions π and σ we say that π *refines* σ if for any block $B \in \pi$ there is a block $C \in \sigma$ such that $B \subset C$. This relation induces an order relation, so we write $\sigma \leq \pi$ if π refines σ.

One crucial element in partition logic is that of defining an analogue for the notion of element of a subset. This is done by introducing the concept of *distinctions* of a partition.

Informally, the distinctions of a partition are the pairs of elements in distinct blocks of the partition. In more formal terms, we say that a pair $(u, u') \in U \times U$ is a *distinction* or *dit* of the partition π if there are distinct blocks $B, B' \in \pi$ with $u \in B$ and $u' \in B$.

The set of distinctions of a partition π, or $dit(\pi) \subseteq U \times U$ is the partition seen as a partition relation:

$$dit(\pi) = \bigcup_{B, B' \in \pi, B \neq B'} B \times B'$$

By the same token, we say that a pair (u, u') is an indistinction or *indit* of a partition π if u and u' belong to the same block of π. The set of indistinctions of π is denoted by *indit(π)* and is just the complement of dit(π) with respect to $U \times U$.

6.4.2 Operations

The *join* of two partitions π and σ is denoted by $\pi \vee \sigma$ and is defined as the partition whose blocks are the non-empty intersections of the blocks $B \cap C$ of the blocks of π and σ.

Example 8 Consider the set $U = \{a, b, c, d, e\}$ and the two partitions

$$\pi = \{\{a, b\}, \{c, d\}, \{e\}\}$$

and

$$\sigma = \{\{a, b, c\}, \{d, e\}\}$$

Then we have

$$\pi \vee \sigma = \{\{a, b\}, \{c\}, \{d, e\}\}$$

The *meet* of two partitions π and σ is denoted by $\pi \wedge \sigma$ and is defined as the finest of the partitions which are coarser than both π and σ.

Example 9 Consider the set $U = \{a, b, c, d, e, f\}$ and the two partitions

$$\pi = \{\{a, b\}, \{c\}, \{d, e\}, \{f\}\}$$

and

$$\sigma = \{\{a\}, \{b, c\}, \{d\}, \{e, f\}\}$$

Then we have

$$\pi \wedge \sigma = \{\{a, b, c\}, \{d, e, f\}\}.$$

The *implication* $\sigma \implies \pi$ is defined as being the partition that has the same structure of π, except that every block $B \in \pi$ which is also contained in some block $C \in \sigma$ appears discretized in $\sigma \implies \pi$.

Example 10 Consider the set $U = \{a, b, c, d, e\}$ and the two partitions

$$\pi = \{\{a, b\}, \{c, d, e\}\}$$

and

$$\sigma = \{\{a, b, c\}, \{d, e\}\}.$$

Then we have

$$(\sigma \implies \pi) = \{\{a\}, \{b\}, \{c, d, e\}\}$$

Finally, given a partition π, the *negation* $\neg\pi$ is defined as $\pi \implies \mathbf{0}$.

6.4.3 Partition Logic and Quantum Mechanics

The brief outline of partition logic given in the previous section is only meant to be a very introductory exposition of a logical system for which many fundamental questions are still open. The most important of such questions in relation to our discussion on realism and iconicity in quantum mechanics concerns the place where partition logic is situated with respect to classical and intuitionistic logic.

Ellerman shows that partition logic is not distributive, and, although it is developed following an intuitionistic flavor, it does not have the needed structure of a Heyting algebra. In order to clarify the real connection with the Boolean logic of subsets, we need to work at the level of categorical logic, where the duality between subsets and partitions becomes evident.

In spite of the aforementioned duality, however, it should be noticed that partitions are much more complex than subsets in many respects. For instance, a Hilbert-style axiom system for partition tautologies together with a proof of completeness is not currently known, and the existence of a finite countermodel for any non-tautology in subset logic is trivial but the corresponding finite model property for partition logic is also an open question [8].

Along with questions concerning subset-partition dualities, Ellerman points out that his logic has an immediate role in representing the notion of *objective indefiniteness* in quantum mechanics (QM) ([8], p. 124):

> Every block $B \in \pi$ in a partition can be interpreted not simply as a subset $B \subseteq U$ of definite elements but as one indefinite element (represented by the collecting together or superposition of its elements) that with further distinctions (measurements in QM) could become any of the fully definite or eigen elements $\{u\}$ for $u \in B$. In this manner, the (non-singleton) blocks of a partition are interpreted as objectively indefinite elements represented as a "superposition" of the fully definite or "eigen" elements that it could become as it is objectively in-formed by more distinction-making "measurements." Seen in this manner, partitional mathematics is just the set version of the mathematical machinery of QM, or, put the other way around, the mathematics of QM can be obtained by "lifting" the machinery of partitions on sets to complex vector spaces.

First of all let us highlight the simplicity of this idea and its analogies with classical QM with an example. Consider a set $U = \{a, b, c\}$ and the partition $\pi = \{\{a\}, \{b, c\}\}$. The block $\{b, c\}$ can be said to be objectively indefinite between b and c, in the precise sense that a *distinction* can result in either b or c to be actually instantiated.

At first glance this may not seem to be any different from saying that a certain quantum state is in the superposition $\alpha|i\rangle + \beta|j\rangle$ where α and β are two complex numbers and i and j two basis complex vectors.

An important philosophical point is at stake, however. Ellerman indeed shows that the correspondence between partitions and vector spaces does provide a complete framework where all the fundamental entities, relations and results of QM can be recast. Hence, he shows that sacrificing the richness of the complex number field for the much weaker structure of sets and partitions does not compromise the overall soundness of the theory.

At the core of his construction, Ellerman places the lifting of set partitions to vector space partitions. Consider a field K and a set U, and associate to U the vector space K^U spanned by the Kronecker delta basis $\delta_u(u') = 0$ if $u \neq u'$, $\delta_u(u) = 1$. This association generates a free vector space functor between the category of sets **Sets** to the category of vector spaces over K.

For instance, if U is the two-element set $U = \{i, j\}$ and $K = \mathbb{R}$, a generic element of K^U has the form $\mathbf{w} = \alpha\delta_i + \beta\delta_j$, with α and β real numbers. Notice that $\mathbf{w}$ is a function defined on U, and $\mathbf{w}(i) = \alpha$, $\mathbf{w}(j) = \beta$. In other words, $\mathbb{R}^{\{i,j\}}$ is in a bijective correspondence with points in the real plane: $\mathbb{R}^{\{i,j\}} = \mathbb{R}^2$. What is really interesting here is that, although we do not have any natural way to "multiply" sets with element of the field K, the free action still allows us to make sense of the idea of a vector space "spanned" by a set.

Ellerman's abductive move then becomes an experimentation about the validity of certain fundamentally non-realistic relations of QM, within the model of QM over sets—in the sense outlined above—using the language of partitions.

There are a certain number of technical details involved in the process of showing that through the development of the QM/sets version of QM results such as Bell's Theorem, the indeterminacy principle, double-slit experiments, and much else can be recovered in the context of finite sets, and we refer the interested reader to [8].

What is more relevant to our presentation is not proving that Ellerman's approach is the correct one—or that it is somehow better than other attempts to produce a new realist interpretation of quantum mechanics. We are indeed interested in the first place in the iconic nature of his original insight within a relational context (categorical duality between partitions and subsets) and, in the second place, in his diagrammatic experimentation efforts underlying the emergence of the most intuitive context in which to represent exactly those quantum mechanical facts that are intrinsically non-realistic.

6.5 Conclusion

Flori, Epperson and Zafiris, and Ellerman present three different ways to reconcile quantum mechanics with a realist orientation in the philosophy of science. For Flori, the anti-realist interpretations grounded in the standard formalism give way to realist reinterpretations when reformulated in the context of topoi. For Epperson and Zafiris, a reconstruction within sheaf-logic and topoi of quantum mechanics very similar to Flori's supports a somewhat more controversial "relational realism" that amounts to a revisionary Whiteheadian metaphysics in which concrete, continuous spatio-temporal event structures are themselves specifications or contractions of more general and fundamental logical/mereotopological algebras as represented categorically by lattices. The relation between the two levels is represented in turn *also* categorically by presheaves and furthermore the functorial lifting of presheaves to sheaves then corresponds to the quantum mechanical passage from pure superpositions of states to calculable probabilities of observable outcomes of measurements. Crucial here is the ontic interpretation of germs. For Ellerman, the logical algebra derived from partitionings of a fixed set allows for the representation of various "paradoxical" QM phenomena in a more familiar (non-complex) mathematical environment, where they may be more interpretatively tractable. Here, it is not a matter of recasting the whole of quantum mechanics in a new mathematical environment but rather of showing how certain specific interpretative aspects of quantum mechanics may be clarified by extracting from the full and completely integrated theory the relevant structural features that determine those aspects *in abstracto*. In particular, the quantum "strangeness" of non-locality and Born's rule are shown by Ellerman to be explicable in terms of partition logic irrespective of any further complications introduced by the use of vector algebras involving complex quantities and other features of standard quantum mechanics.

In each case, a realist philosophical orientation comes neither by fiat nor by simple reinterpretation of already given mathematical formalisms using revised ontic or ontological concepts. Instead, the reformulation of the familiar mathematical structures in new mathematical contexts gives rise to new concepts in each of the three distinct projects through the very shift in perspective that each particular reformulation entails. In Flori, the pre-established framework of realism is shown to be *preserved* when the otherwise non-realist coherent histories interpretation is formulated in topoi via sieves. In Epperson and Zafiris, essentially the same mathematical reformulation as that of Flori results instead in a revised concept of the ontology that underlies scientific realism itself. The building-blocks of the real are no longer discrete objects or points of space-time, but rather locally synthetic tendencies corresponding to germs in sheaf-theory. Finally in Epperson, partition logic provides a heuristically useful mathematical framework for isolating just those aspects of quantum theory that might appear anti-realist and "paradoxical" but are in fact perfectly comprehensible when examined from the formally appropriate point of view.

We are intending neither to support nor to critique a realist reinterpretation of quantum mechanics in general. Nor do we aim to assess the relative merits of the three approaches discussed above. Our concern is to treat this trio as a set of illustrations of abductive reasoning that uses mathematics. From an abductive point of view, what is at stake is emphatically *not* whether or not these distinct albeit related realist interpretations of quantum mechanics are in fact correct, that is, whether or not the ultimate ontic particulars of our universe are point-like or germ-like or if quantum state-superpositions are really somehow block-like entities that are susceptible of partitional refinement. These are highly interesting questions, but they are not ours here. What matters instead for the present inquiry is the kind of reasoning that is implied in Flori, Epperson and Zafiris, and Ellerman being led to make such global claims about quantum mechanics at all, since quantum mechanics as traditionally formulated suggests a *prima facie* anti-realist interpretation.

What Flori, Epperson and Zafiris and Ellerman collectively show, in our view, is that the most general theoretical context in which philosophical reflection on quantum mechanics takes place is not only a matter of interpreting the mapping of mathematics to the real, but includes variability (and abduction) within mathematical representation itself. The question of scientific realism is a question of the relation between scientific theories and their mathematically iconic tools on the one hand and ontic-ontological reality on the other. To put this relation itself genuinely into question is to suspend any fixed conception of what either of its terms is independently.

References

1. F. David, *The Formalisms of Quantum Mechanics* (Springer, Heidelberg, 2015)
2. C. Heunen, M. Sadrzadeh, E. Grefenstette, *Quantum Physics and Linguistics: A Compositional, Diagrammatic Discourse* (Oxford University Press, Oxford, 2013)
3. C. Flori, A First Course in Topos Quantum Theory, Lecture Notes in Physics, **868** (Springer, Heidelberg, 2013)

4. A. Döring, C.J. Isham, What is a thing?: Topos theory in the foundations of physics. Lect. Notes Phys. **813**, 753–937 (2010)
5. M. Epperson, E. Zafiris, *Foundations of Relational Realism: A Topological Approach to Quantum Mechanics and the Philosophy of Nature* (Lexington Books, Lanham, 2013)
6. A.N. Whitehead, *Process and Reality* (Basic Books, New York, 1929)
7. S. Mac, L. Moerdijk, *Sheaves in Geometry and Logic: A First Introduction to Topos Theory, (Universitext)* (Springer, Heidelberg, 1994)
8. D. Ellerman, An introduction to partition logic. Log. J. IGPL **22**(1), 94–125 (2014)

Conclusion

The kind of thinking associated with category theory is geometrical and diagrammatic in essence, based as it is on formal relations of embedding and variation that hold between "figures" and "spaces" definable in very general terms. As we have seen in particular in chapter six, the role of spaces may be played for instance by ambient mathematical environments, and the corresponding figures are then the formal structures of quantum mechanics that are given as a heterogenous collection of data through the histories of experiment and theorization in modern physics. As these structural figures are embedded in the diverse mathematical spaces opened by category theory, topoi and partition logic, they become available to philosophical reflection as *diagrams* subject to experimental control and investigation. The concepts that in turn must be used to interpret these diagrams are then not given arbitrarily, but must instead be developed pragmatically through collective processes of experimentation and revision that are themselves scientific in character and are thereby available to critical scrutiny.

This way of thinking about scientific models opens up the possibility of a kind of realism with respect to processes of scientific modelization through mathematics themselves. This realist orientation with respect to epistemological method arises as a theoretical result, not a mere presupposition or posit. The result in question in any given case, however, is by no means a deductive certainty. Yet neither is it an inductive generalization or calculable probability based on given data. Instead, the reasoning at work here takes the shape of abductive inference. The "surprising event" initiating the abductive process may be, for example, the concrete history of twentieth and twenty-first century subatomic physics as both a constellation of real events and a sequence of epistemological forays, detours and extensions.

Much philosophy of science remains trapped in simplistically dualist conceptual frameworks, even and especially where a traditional Cartesian metaphysical dualism of extended and thinking substances is denied. The more insidious dualisms arise through the exportation of implicit meta-theoretical distinctions directly into meta-theoretical analysis and reflective discourse. It is obvious, for example, that the expression of most scientific theories is spontaneously "realist" in the straightfor-

G. Caterina and R. Gangle, *Iconicity and Abduction*,
Studies in Applied Philosophy, Epistemology and Rational Ethics 29,
DOI 10.1007/978-3-319-44245-7

ward and naive sense that they are ostensibly theories "about" real objects and their various properties and relations. Some philosophers of science, however, seem to take this contingent fact as a fundamental ground for meta-theoretical reflection. Yet if this lifting of spontaneous meta-theory to philosophically reflective meta-theory is justifiable, it ought to be tested and justified explicitly. After all, we take it for granted in a modern scientific context that scientific theories may come to challenge our ordinary or "folk" theories about the natural world (Copernican astronomy, relativity, quantum mechanics, etc.). This is what scientific inquiry *essentially* does. Our "folk" conceptual frameworks regarding science need to be equally subject to critical inquiry and revision. No doubt when formulated in this way, nearly all philosophers of science would be inclined to agree. Such agreement, however, might be more realized in principle than in practice. Philosophers tend to forget that the iconic and diagrammatic models that characterize scientific thinking are just that: models. For instance, while formal languages often serve as useful models of scientific theories, scientific theories themselves simply *are not* sets of statements in formal languages. We may not know exactly what they are, any more than we know exactly what lightning is or gravity prior to scientific inquiry. But just because they are theoretical in character and employ models does not imply that they are equivalent to their own theoretical modelizations. They are, rather, always iconic expressions of their own epistemological trajectories through a partly obscure space of knowledge, ignorance and experience. Thus, not just the truth or falsity of scientific realism but the question of *what realism even means* ought to be a matter of sustained inquiry.

The earlier analyses of Peircean semiotics and logical grammar and Badiou's mathematical models of ontology and epistemology may be understood in the light of this notion of a scientific realism with respect to processes themselves of scientific investigation. For both of these philosophers, the formal models of knowledge and being that track the scientific investigation of nature and its essential use of mathematics are themselves expressions of ongoing abductive processes supported by complex cultural and historical dynamics at multiple levels. Such a view does not imply epistemological relativism, but rather emphasizes the context-bound reality of all actual movements of scientific inquiry. Abductive reasoning is a crucial aspect of all scientific modes of knowing, and it operates at both macroscopic and microscopic scales of inquiry. Continued investigation into the dynamics of abductive inference both as concretely evident in actual science and as formally guiding new procedures of optimized hypothesis construction remains one of the most important elements of philosophy of science and indeed of science itself.

Appendix A
Category Theory: A Primer

This appendix is meant only to fix some elementary notation and terminology for the uses made of category theory in the main text and to provide some motivating examples for those who might be unfamiliar with the material. Comprehensive surveys of this mathematical field must be sought elsewhere. For more complete treatments of elementary category theory see [1]. An excellent introduction for those with no background whatsoever in the subject is found in Lawvere [2].

A.1 Category Theory Axioms

A *category* is an abstract system of objects and relations. Categories are readily imagined by diagrams of dots and arrows. Objects are differentiated as distinct nodes in the system, whereas the relations link one individual to another in some determinate way. Between any two dots, there may be many arrows, one or none, and there may be arrows from any given object to itself (indeed every object must have at least one of these).

G. Caterina and R. Gangle, *Iconicity and Abduction*,
Studies in Applied Philosophy, Epistemology and Rational Ethics 29,
DOI 10.1007/978-3-319-44245-7

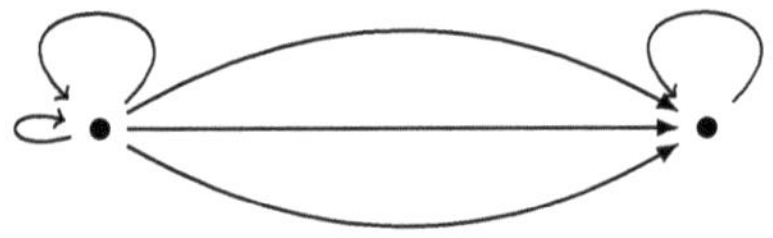

Arrows are then subject to the following three axioms, which all categories must satisfy and which are together sufficient to determine a category as such:

(A1) Axiom of Composition: Given any two arrows ordered such that the target of one is the source of the other, the composition of the first followed by the second exists in the category as a unique and definite arrow.

Formally, given two arrows

$$A \xrightarrow{f} B$$

$$B \xrightarrow{g} C$$

there exists one and only one arrow

$$A \xrightarrow{g \circ f} C$$

This arrow $g \circ f$ is called the *composite* of f and g. This "composition-relation" among arrows is intuitively captured by the following diagram:

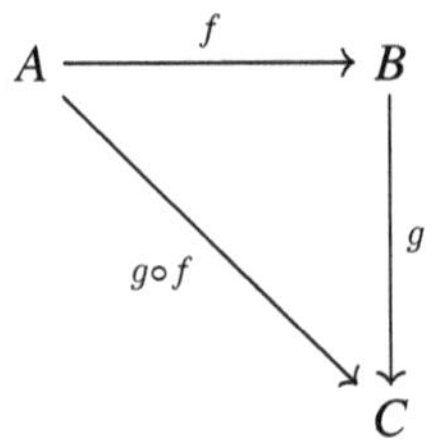

Remark 1 It is important to realize that the structure of a category is defined essentially in terms of how its arrows compose. A useful analogy here can be made with

group theory: typically, the structure of a group may be exhaustively expressed by the relations among its generators.

In order to highlight this fact, it is useful to look at the two pictures below. They both appear to have the same structure of objects and arrows, but the stipulated arrow-relations are different, therefore leading to two distinct mathematical entities.

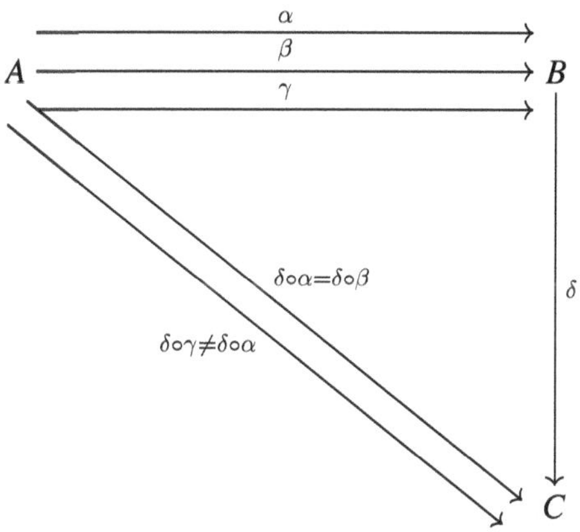

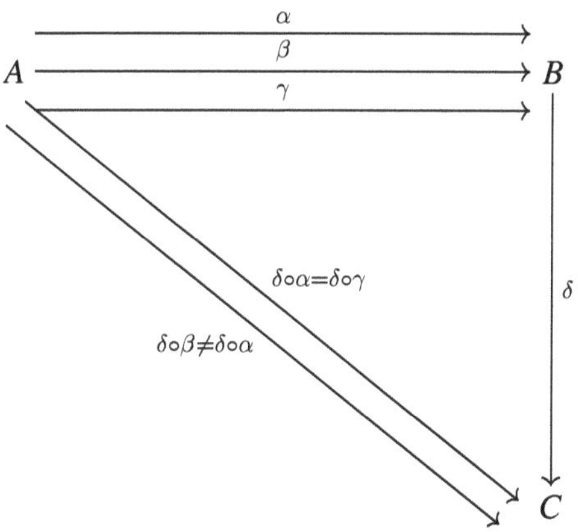

(A2) Axiom of Associativity: Given any three arrows ordered such that the target of the first is the source of the second and the target of the second is the source of the third, the composite of the first two arrows composed with the third is the same as the first arrow composed with the composite of the second and third.

Formally, given three arrows

$$A \xrightarrow{f} B$$

$$B \xrightarrow{g} C$$

$$C \xrightarrow{h} D$$

the following diagram *commutes*, that is, all diagram arrow-paths with common sources and targets collapse to identical arrows (see below for a fuller explanation of this key diagrammatic notion):

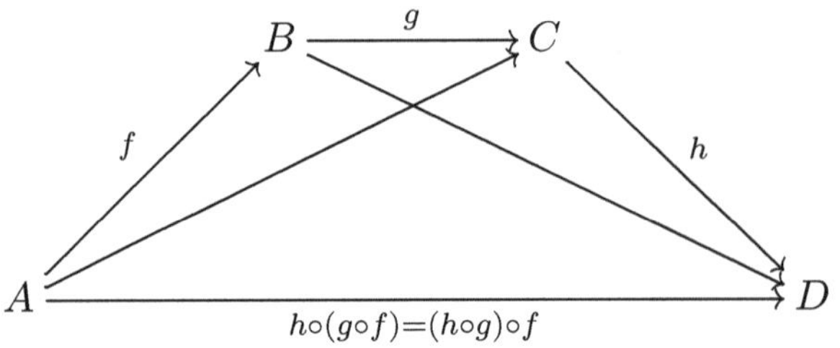

In practice, this means that any path of arrows in a category composes to a unique arrow.

(A3) Axiom of Identity: Every object has an arrow called its *identity* (Id) which takes that object as both its source and its target and which composes "inertly" with all arrows having a composition defined with it. See the diagram and compositional equation immediately below:

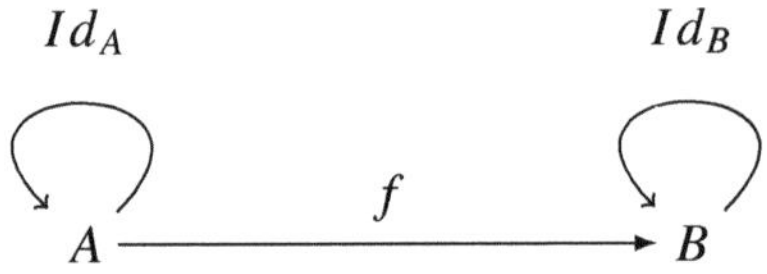

$$f = f \circ Id_A = Id_B \circ f,$$

for all arrows f from A to B.

In practice, this means that identity-arrows may be arbitrarily added to or subtracted from paths of arrows without altering the composition of the path. In addition, since each identity arrow uniquely picks out the object to which it is attached and represents the static and inert iteration of that object's identity, from a categorical standpoint each object may typically be "identified" with its identity arrow.

In fact, it is possible to define categories in an "arrows-only" fashion without reference to objects at all. As [3] in particular emphasizes (p. 261),

> the usual distinction between objects and morphisms of categories is formally dispensable: since with each object A of given category C is associated a unique identity morphism 1_A, one may formally identify objects of C with their corresponding identity morphisms and thus consider objects as morphisms of special sort. Thus, formally, a general category can be described as a class of things called morphisms provided with a (partial) binary associative operation called composition.

A.2 Diagrams

Informally, a diagram in a category is a collection of dots (or labels) representing objects connected by arrows representing arrows (note that in this way, arrows in a category theory diagram interestingly "represent themselves"). We may speak of the system of objects and arrows as *represented by* such a diagram as being that diagram *in* the category at issue, whereas the (same) diagram conceived as *representing* this system may be said to be a diagram *for* this same category.

A diagram is said to *commute* when all directed paths in the diagram with the same start and endpoints lead to the same result by composition. For instance, if we declare that the following diagram commutes:

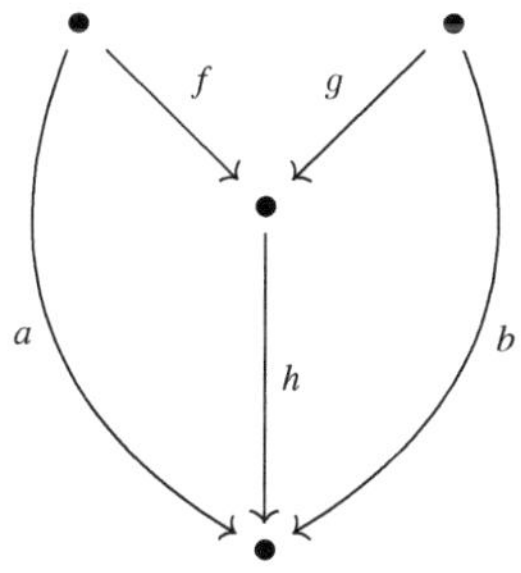

we are really declaring that the arrow a *is and must be* the composition $h \circ f$ and the arrow b *is and must be* the composition $h \circ g$.

In this sense, when we say that a diagram is commutative we are in fact imposing certain relations on the variable components (arrows) of our structure (a category). This is similar to imposing relational identities (equations) on structures based on "classical" symbolic variables (think, for instance, of the equation $xy = yx$ which distinguishes abelian groups amongst all groups).

We want to highlight in this way that the notion of commutative diagram expresses the same "intuitionistic" problem intrinsic to the concept of equality in mathematics more generally: How can two ostensibly different terms, for instance in an algebraic expression, "come to be" conceived as representing in fact one and the same "thing"? By looking at the diagram above we see two arrow-paths $h \circ f$ and a, therefore two distinct objects, that are actually meant to represent (and are in fact forced to be) the same mathematical entity on the basis of the stipulated commutativity of the diagram taken as a whole.

Furthermore, when defining structures within a category, a variety of formal notational devices may be used to handle issues of quantification ("*for all arrows* f", "*there exists exactly one arrow* g", etc.). See, for instance, the innovative diagrammatic approach proposed in [4]. In practice, cases involving significant ambiguity seldom arise.

A.3 Isomorphisms

Two objects A and B are said to be *isomorphic* if there exist two arrows f and g as given below such that the following diagram commutes.

$$Id_A \circlearrowleft A \underset{f}{\overset{g}{\rightleftarrows}} B \circlearrowright Id_B$$

Informally, since this structure ensures that the objects A and B have identical systems of arrow-relations with all other objects in the ambient category, from a strictly category theoretical perspective they may be treated *in the context of this particular category* as being "the same".

A.4 Functors Between Categories

Categories may be related to one another via mappings called *functors*, which may be understood at a first approach on analogy with functions between sets.

Given two categories $\mathcal{C}$ and $\mathcal{D}$, a functor is in the first place a map F from objects of $\mathcal{C}$ into objects of $\mathcal{D}$ and from arrows of $\mathcal{C}$ into arrows of $\mathcal{D}$, that is, roughly, a function from the set of the objects of $\mathcal{C}$ to the set of the objects of $\mathcal{D}$ together with a function from the set of arrows of $\mathcal{C}$ to the set of arrows of $\mathcal{D}$.[1]

The condition that this map has to satisfy is that (a) relations linking arrows to their source and target objects are preserved and (b) composition relations between arrows are preserved across the mapping.

This amounts to saying that, if

$$A \xrightarrow{f} B$$

is an arrow in $\mathcal{C}$, then

$$F(A) \xrightarrow{F(f)} F(B)$$

is an arrow in $\mathcal{D}$.

Furthermore, the following diagram must commute in $\mathcal{D}$ for all suitable objects A, B and C and arrows f, g and $g \circ f$ in $\mathcal{C}$:

[1] This formulation only causes difficulties in the (not infrequent) cases when the objects and/or arrows of either $\mathcal{C}$ or $\mathcal{D}$ cannot be gathered into a set, for instance when one of these is the category **Sets** of sets and functions. The ensuing problems and the various strategies for resolving them are readily located in the standard literature on categories.

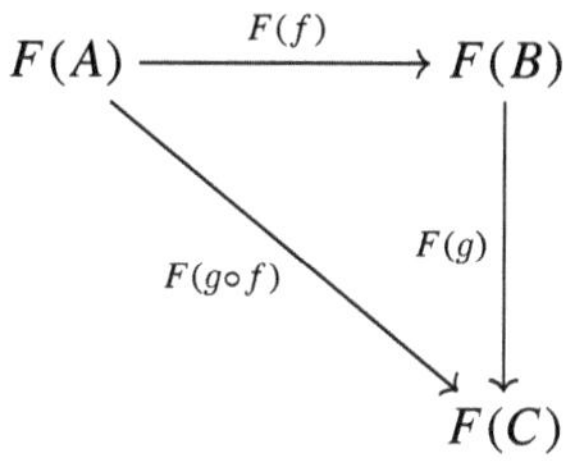

Finally, identity arrows must "track" with their objects across the mapping. Formally, for any object A in $\mathcal{C}$,

$$F(Id_A) = Id_{F(A)}$$

Such a mapping F is called a *covariant functor*. There is a dual notion of *contravariant functor*, which, essentially, instead of "preserving" arrows it "reverses" them. That is, $F : \mathcal{C} \longrightarrow \mathcal{D}$ is said to be *contravariant* if, given two object A and B in $\mathcal{C}$ we have[2] that

$$(f : A \longrightarrow B) \Rightarrow F(f) : F(B) \longrightarrow F(A)$$ [2]

The reader familiar with abstract algebra will appreciate the striking similarity between the notion of functor and that of homomorphism in algebra. Indeed, if we think of category theory as an algebra whose elements are arrows and where defining equational structures are represented by commutative diagrams, the above diagram (for the covariant case) simply says that $F(g \circ f) = F(g) \circ F(f)$.

A.5 Diagrams as Functors

With the notion of functor in hand, it becomes possible to define categorical diagrams in a formal way.

Following the treatment given in [1], given two categories $\mathcal{J}$ and $\mathcal{C}$, we define a *diagram of type* $\mathcal{J}$ in $\mathcal{C}$ to be a functor

$$D : \mathcal{J} \to \mathcal{C}$$

The image of the functor D may then be said to be a diagram of $\mathcal{J}$ in $\mathcal{C}$. In particular, if we think of $\mathcal{J}$ as the index category and write its objects using the letters $i, j, \ldots$, we can then indicate the values of such functors with the symbols $D_i, D_j, \ldots$ and similarly with arrows.

The key idea here is to think of the diagram D as a "picture of $\mathcal{J}$ in $\mathcal{C}$."

---

[2]Contravariant functors also reverse the direction of composition.

A.6 Limits

Among the diverse species of mathematical constructions possible in category theory, *limits* and *colimits* (dual to limits) are at once fundamental to and deeply iconic of the essential features of the categorical approach. The presentation given here follows in broad strokes that of [1], pp. 101–104.

Every basic introduction to category theory includes the definition of certain "special" objects that may or may not exist in a given category, such as "the product of two objects", "equalizer", "pullback", "exponential object", "subobject classifier", just to mention some of the most fundamental. These objects may be defined in terms of certain "universal properties" of types of diagrams in whatever category is in question (in the formal sense of diagram defined above).

For instance, given two objects A and B of a category $\mathcal{C}$, we can interpret the commutative diagram below

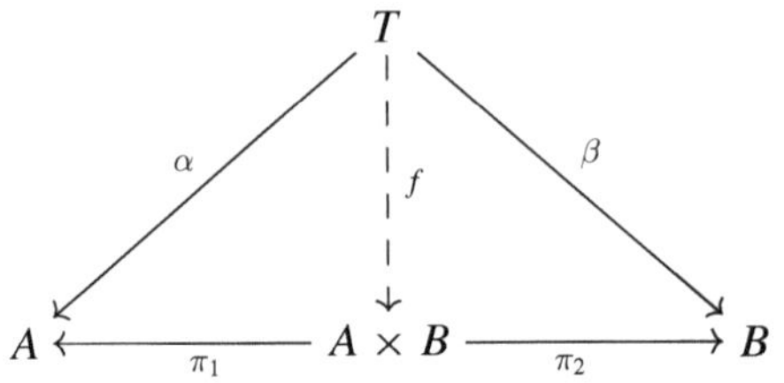

as saying that the categorical product $A \times B$ is an object such that together with the two morphisms π_1 and π_2 from $A \times B$ into A and B, respectively, given *any* test object T in $\mathcal{C}$ and *any* two morphisms α and β from T to A and B respectively, there exists one and only one morphism from T to $A \times B$ (represented by the dashed arrow f) which makes the diagram as a whole commute. This construction determines *products* in the categorical sense as a particular type of *limit*.

Two main ingredients play a fundamental role in "unpacking" categorical diagrams as limits:

(1) a "basic shape", such as

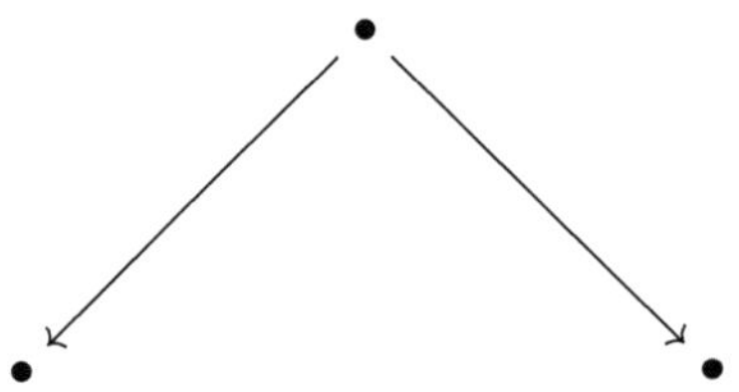

and

(2) a *universal property* to be satisfied.

An elegant and very unifying approach to the understanding of categorical constructions is to formalize these two concepts (basic shape and universal property) through the notion of limits of diagrams, in the precise sense defined in the previous section.

Using the notation introduced above, we define a *cone* to a diagram D in $\mathcal{C}$ as consisting of an object $C \in \mathcal{C}$ and a family of arrows

$$c_j : C \longrightarrow D_j$$

one for each object $j \in \mathcal{J}$, such that for each arrow $\alpha : i \longrightarrow j$ in $\mathcal{J}$ the following diagram commutes:

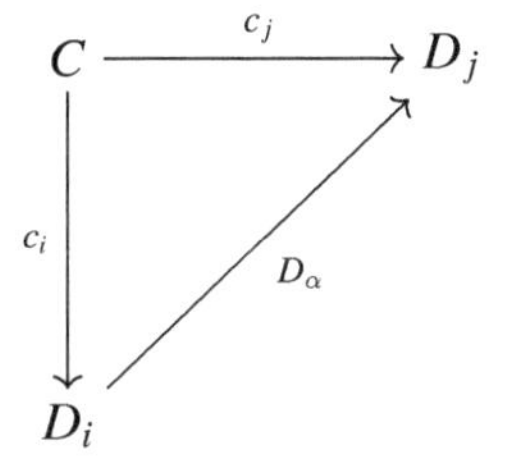

Quoting Awodey [1] "A cone to such a diagram D is then imagined as a many-sided pyramid over the 'base' D and a morphism of cones is an arrow between the apexes of such pyramids".

The collection of all cones to D is denoted by **Cone**(D) and this collection can be shown to form a category in its own right once an opportune, natural notion of arrow, or morphism, between these objects is defined.[3]

We now define the straightforward notion of *terminal object* in a category and then use this construction in order to complete our general formulation of categorical limits as universal properties of types of diagrams.

Definition 10 In a category C an object A is a *terminal object* iff for any object T of C there is exactly one arrow $T \longrightarrow A$.

Definition 11 A *limit* for a diagram $D : \mathcal{J} \longrightarrow \mathcal{C}$ is a terminal object in **Cone**(D). A *finite limit* is a limit for a diagram on a finite index category $\mathcal{J}$.

These somewhat abstract definitions may be clarified through a simple example.

Example 11 (*Awodey*) Take $\mathcal{J} = \{1, 2\}$ the discrete category with two objects and no non-identity arrows. A diagram $\mathcal{D} : \mathcal{J} \longrightarrow \mathcal{C}$ is a pair of objects $D_1, D_2 \in \mathcal{C}$. A cone on D is an object of $\mathcal{C}$ equipped with arrows

$$D_1 \xleftarrow{\quad c_1 \quad} C \xrightarrow{\quad c_2 \quad} D_2$$

And a limit of D is a terminal such cone, that is, a product $D_1 \times D_2$ in $\mathcal{C}$ of D_1 and D_2.

[3] The intuitive notion of an arrow from one cone to another should be clear (essentially, it is a arrow from C to D in $\mathcal{C}$ such that "everything commutes"). The technical details may be found in [1], p. 102.

The reader should compare this last example with the diagram given at the beginning of the present section to understand the utility and economy of this formal approach.

A.7 Examples of Categories

We are especially interested in how the type/token relation is typically presented categorically. Many mathematical "types" naturally constitute categories in which the objects of the category are all of the tokens of the given type and the morphisms are maps between such objects that preserve their structure in some relevant way. Note that this effectively defines a given categorical "type" as a coherent system of relations. In many cases (such as in all of the examples provided below) this system of "external" relations among the tokens is sufficient to determine all of their relevant "internal" structure.

In addition, it is often the case that each of the tokens of a standard mathematical type may be represented by a category. This is the case for all the examples discussed below. Thus from this perspective, category theory may serve as a mathematical tool that in contexts like those illustrated by the following examples treats mathematical types and tokens from within a common framework.

- A *set* S is an unstructured collection of elements.
  - The category **Sets** has for objects sets $R, S, T \dots$ and for arrows functions $f : R \longrightarrow S$ between sets, for all sets R and S. In this category isomorphisms are bijections, thus isomorphic objects correspond to sets of the same cardinality.
  - Every individual set S may also be conceived as a category $\mathcal{C}_S$. The objects of $\mathcal{C}_S$ are the elements $e, f, \dots$ of S and the only arrows in $\mathcal{C}_S$ are the identity arrows. This is sometimes called the *discrete category* on S.
- A *group* G is a set endowed with an associative binary operation $\star$, such that: (a) there is an element $e \in G$ with the property that $e \star g = g = g \star e$ for all $g \in G$; (b) for any $g \in G$ there is an element g^{-1} such that $g \star g^{-1} = g^{-1} \star g = e$.
  - The category **Groups** has for objects groups and for arrows all group homomorphisms between groups.
  - Every individual group G may also be conceived as a category. The abstract structure of G is given by a category $\mathcal{C}_G$ with exactly one object and all of the arrows of which are invertible, that is, for every arrow f there is an arrow f^{-1} such that $f \circ f^{-1}$ is the identity arrow on the (only) object of $\mathcal{C}_G$. Arrow composition then corresponds to the group operation given by G. Functors from $\mathcal{C}_G$ into **Sets** select permutation groups on the elements of the individual sets selected by the mappings of the one object in $\mathcal{C}_G$.
- A *groupoid* is an algebraic generalization of a group in which the binary relation $\star$ is replaced by a partial function. For details see [5].

  - The category **Groupoids** has groupoids for objects and groupoid homomorphisms for arrows.
  - Each groupoid may be conceived straightforwardly as a category all of whose arrows are invertible. Intuitively, these are categories that are structured like groups but which may have more than one object.

- A *partial order* on a set P is a binary relation $\leq$ defined on P that is reflexive, antisymmetric and transitive.

  - The category **ParOrders** has partial orders for objects and monotonic functions between partial orders (functions that preserve the order relation $\leq$) as arrows.
  - Every individual partial order $P, \leq$ may also be conceived as a category $\mathcal{C}_{P,\leq}$. The objects of $\mathcal{C}_{P,\leq}$ are the elements of P and given any two elements P_p and P_q of P, an arrow $a : P_p \longrightarrow P_q$ exists in $\mathcal{C}_{P,\leq}$ if and only if $P_p \leq P_q$.[4]

- Partial orders may be generalized to *preorders* by relaxing the antisymmetry condition of partial orders. In other words, a preorder is a set P with a binary relation $\leq$ defined on P that is reflexive and transitive.

  - The category **PreOrders** has preorders for objects and order-preserving functions between preorders as arrows.
  - Each preorder $P, \leq$ determines a category defined in exactly the same way as a partial order above, although of course without the relation $\leq$ having to be antisymmetric. Alternately, preorder categories may be characterized simply as categories such that there is at most one arrow in either direction between any two objects. Note that this implies in particular that for such a category, A and B are necessarily isomorphic if arrows exist $f : A \longrightarrow B$ and $g : B \longrightarrow A$.

  The reader should recognize how the generalization of groups to groupoids is strictly analogous to that of partial orders to preorders. Furthermore, the reader should note how the respective categorical characterizations make this analogy particularly evident.
- Finally, we introduce a somewhat unorthodox category $\mathcal{R}_W$ of dyadic relations defined on some set I of individuals in a world W. The objects of the category are these individuals themselves, that is, elements of I, and morphisms in the category are all dyadic relations instantiated in the world W between all pairs of such individuals. We presume that every individual has the relation "is identical to" to itself, and we take advantage of the fact that dyadic relations compose naturally in a canonical way, namely by concatenation. More precisely, given three individuals A, B and C and two dyadic relations $r_1 : A \longrightarrow B$ and $r_2 : B \longrightarrow C$, a dyadic relation $r_{1,2} : A \longrightarrow C$ is determined as a relation between A and C that concatenates r_2 following r_1. The concatenated relation is "generic" in the sense that the mediating object/individual is replaced by an indeterminate "something". For instance, if r_1 is the relation "is next to" and r_2 is the relation "is father of" then $r_{1,2}$ is the relation "is next to something that is the father of". The reader should

[4]This means in particular that every topological space induces a partial order category whose objects are the open sets and with morphisms corresponding to inclusion maps.

check how the categorical axioms of identity, composition and associativity are satisfied by this characterization. The logical and philosophical interest of this category rests in how the *incidence relations* of the arrows are conceived and stipulated for various semantic and metaphysical views of what relations are and how they work.

A.8 Functor Categories

A special and highly important categorical type consists of categories the objects of which are *functors* and the arrows of which are mappings between functors called *natural transformations*. More precisely, given two categories $\mathcal{C}$ and $\mathcal{D}$, the functor category $\mathcal{D}^{\mathcal{C}}$ is be defined as follows:

- The objects of $\mathcal{D}^{\mathcal{C}}$ are all *functors* $\mathcal{C} \rightarrow \mathcal{D}$.
- The arrows of $\mathcal{D}^{\mathcal{C}}$ are all *natural transformations* between functors $\mathcal{C} \rightarrow \mathcal{D}$.

Natural transformations are morphisms between functors: given two functors $F \in \mathcal{D}^{\mathcal{C}}$ and $G \in \mathcal{D}^{\mathcal{C}}$, a natural transformation between F and G is a family of morphisms η_O parametrized by the objects $O \in \mathcal{C}$ such that the following diagram commutes for any two objects A and B that are connected by a morphism f in $\mathcal{C}$:

$$\begin{array}{ccc} F(A) & \xrightarrow{F(f)} & F(B) \\ \downarrow{\scriptstyle \eta_A} & & \downarrow{\scriptstyle \eta_B} \\ G(A) & \xrightarrow{G(f)} & G(B) \end{array}$$

A.9 Yoneda's Lemma Proof

We follow here the presentation given in [6]. Let us restrict our previous constructions by setting $\mathcal{D} = \mathbf{Sets}$ (with set functions as morphisms). In other words, let us consider the functor category $\mathbf{Sets}^{\mathcal{C}}$ where objects are functors from an arbitrary category into the category of sets.[5]

We first define a canonical functor associated with every object of $\mathcal{C}$.

Definition 12 Given any object $A \in \mathcal{C}$, the *hom-functor based at* A is denoted by $Hom(A, -) : \mathcal{C} \longrightarrow \mathbf{Sets}$ and takes any object $X \in \mathcal{C}$ into the set $Hom(A, X)$ of all the morphisms from A to X in $\mathcal{C}$. How arrows in $\mathcal{C}$ are treated by this functor is discussed below.

[5]For the rest of this section, unless otherwise specified, functors will be understood as belonging to $\mathbf{Sets}^{\mathcal{C}}$.

Example 12

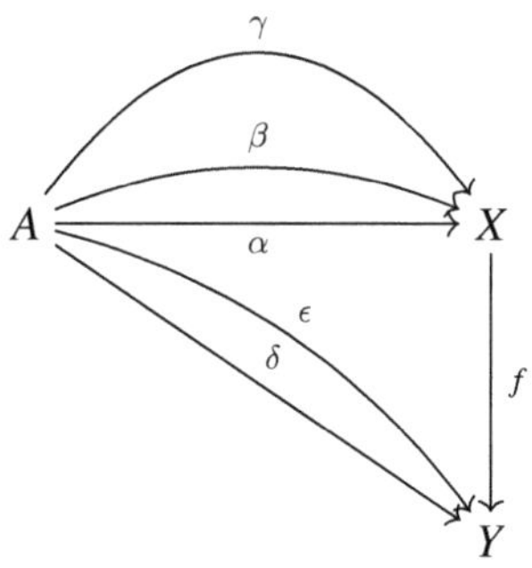

Let us assume the following structural equation for this diagram:

$$f \circ \alpha = \epsilon, \; f \circ \beta = \epsilon, \; f \circ \gamma = \delta$$

We then have

$$Hom(A, X) = \{\alpha, \beta, \gamma\}, \; Hom(A, Y) = \{\delta, \epsilon\}$$

Notice that so far we have only expressed the action of $Hom(A, -)$ on objects. It is an easy exercise to show that $Hom(A, -)$ is an actual functor by acting on morphisms by pre-composition.

In the example above, for instance, $Hom(A, f)$ is a map between the sets $Hom(A, X)$ and $Hom(A, Y)$ defined as follows:

$$Hom(A, f)(\alpha) = f \circ \alpha = \epsilon$$

$$Hom(A, f)(\beta) = f \circ \beta = \epsilon$$

$$Hom(A, f)(\gamma) = f \circ \gamma = \delta$$

Definition 13 A functor F for which there exists an object A such that $F(X) = Hom(A, X)$ for any object $X \in \mathcal{C}$ is said to be *representable*.

It is not quite true that all functors from $\mathcal{C}$ into **Sets** are representable. Yoneda's lemma, however, proves that all such functors can be obtained from hom-functors through natural transformations, and it explicitly parametrizes all such transformations:

Theorem 9 *There is a one-to-one correspondence between natural transformations from $Hom(A, -)$ to F and elements of $F(A)$.*

Proof Consider two objects A and B in $\mathcal{C}$ connected by a morphism f:

$$A \xrightarrow{f} B$$

The naturality square for pairs of functors F and $Hom(A, -)$ is the following:

$$\begin{array}{ccc} Hom(A,X) & \xrightarrow{Hom(A,f)} & Hom(A,Y) \\ \Big\downarrow{\scriptstyle \eta_X} & & \Big\downarrow{\scriptstyle \eta_Y} \\ F(X) & \xrightarrow{F(f)} & F(Y) \end{array}$$

Consider now an element h of $Hom(A, X)$ and let us look at the equation expressed by the commutativity of the above diagram. By remembering that $Hom(A, f)$ acts by pre-compositions, we obtain

$$\eta_Y(f \circ h) = F(f)[\eta_X(h)]$$

By specializing this construction to the case $X = A$, we obtain the following naturality square:

$$\begin{array}{ccc} Hom(A,A) & \xrightarrow{Hom(A,f)} & Hom(A,Y) \\ \Big\downarrow{\scriptstyle \eta_A} & & \Big\downarrow{\scriptstyle \eta_Y} \\ F(X) & \xrightarrow{F(f)} & F(Y) \end{array}$$

In this case h is a morphism from A into A. Since we know that id_A is for sure one of those, we have the following relation:

$$\eta_Y(f) = F(f)[\eta_A(id_A)]$$

Notice the implication of this result: the image of η_Y of any morphism $f \in Hom(A, X)$ is completely determined by assigning a value to η_A at the morphism id_A, that is, any choice of such value will determine a natural transformation.

Conversely, given any natural transformation η, this result tells us that we can evaluate η at id_A to obtain a point in $F(A)$. □

A.10 Further Developments

Higher-level categorical constructions based on Yoneda's lemma open up a vast and still largely uncharted domain of mathematics. In particular, the basic approach to mathematics as grounded in functor categories allows for a purely relational characterization of many established results that frequently provides new insights into their applicability in other mathematical areas. We refer the reader in particular to the intermediate-level text [7] for indications of possible directions for further exploration.

References

1. S. Awodey, *Category Theory*, 2nd edn. (Oxford, 2010)
2. W. Lawvere, Conceptual Mathematics: A First Introduction to Categories (Cambridge University Press, Cambridge, 2009)
3. A. Rodin, *Axiomatic Method and Category Theory* (Springer, Heidelberg, 2014)
4. P. Freyd, A. Scedrov, *Categories Allegories* (Elsevier, Amsterdam, 1990)
5. R. Brown, *Topology: A Geometric Account of General Topology Homotopy Types and the Fundamental Groupoid* (Ellis Horwood Press, Denmark, 1988)
6. M. Bartosz, http://bartoszmilewski.com/2015/09/01/the-yoneda-lemma/
7. M. Reyes, G. Reyes, H. Zolfaghari, *Generic Figures and Their Glueings* (Polimetrica, Milan, 2004)

Appendix B
Topos Theory: A Primer

Topos theory is a particular mathematical genre in which, among other things, the ontological and constructive aspects of abduction may be modeled. The core idea of a topos is that it is a type of category with sufficient resources to build up the objects and operations characterizing ordinary mathematics. Any topos is thus sufficiently rich in objects and relations to construct "meta-objects" (such as products $A \times B$) given any relevant constellation of objects within the topos (such as any pair of objects A and B). The most important point is that the topos is given as a whole: All the meta-objects by which it is structured are—from the internal perspective of the topos itself—simply additional objects. Thus the material out of which ordinary mathematics may be constructed within a topos is the fabric of structural relations that compose the topos itself as a unified and internally differentiated whole.

Topoi are essentially categories possessing a sufficiently rich internal structure of objects and relations to model many of the most important core mathematical and logical constructions needed in ordinary mathematical discourse. A particularly canonical topos and one that will help the uninitiated reader find his or her bearings throughout the following categorical characterization is the topos **Sets** of sets and functions. A topos may in many cases be thought of as a category that is set-like in certain determinate ways.

Topos theory constitutes a vast ocean of mathematical research and insight. Helpful introductions may be found in [1–3]. An aging but still highly relevant attempt to provide a comprehensive survey of topos theoretical results is found in [4]. What follows merely outlines the fundamental axioms of elementary topoi, with a special emphasis on explaining the precise sense in which topoi grasp the essential structure of the category of sets.

G. Caterina and R. Gangle, *Iconicity and Abduction*,
Studies in Applied Philosophy, Epistemology and Rational Ethics 29,
DOI 10.1007/978-3-319-44245-7

B.1 Structure of Sets

In this section we first recall some structural features of sets, using a set-theoretical language, with the goal of framing them in a categorical (hence relational) context. The important point in each of the following constructions is that a familiar property of sets and functions may be characterized in the category **Sets** solely in terms of relational properties of arrows. By then relying *only* on these categorical characterizations, a greatly enriched space of "set-like" categories emerges.

B.1.1 *Injective Functions*

A function f between two sets A and B is said to be *injective* if $(f(x) = f(y)) \Rightarrow x = y$.

The categorical translation of this property in **Sets** is given by the notion of monic arrow: Given

$$A \xrightarrow{\ f\ } B$$

f is said to be *monic* if, for any two arrows

$$Z \underset{g}{\overset{h}{\rightrightarrows}} A$$

such that the pair of compositions represented by the diagram

$$Z \underset{g}{\overset{h}{\rightrightarrows}} A \xrightarrow{\ f\ } B$$

are equal (that is, such that $f \circ g = f \circ h$), it must be the case that

$$h = g$$

It can be shown that, in the category **Sets**, every monic arrow corresponds to an injective function.

B.1.2 *Pullbacks*

Given two functions with the same codomain and possibly two different domains, say

$$f : A \longrightarrow C$$

and

$$g : B \longrightarrow C,$$

we can form the set $A \times_C B$ constituted by the ordered pairs (a, b), with $a \in A$ and $b \in B$, such that $f(a) = g(b)$.

Let us now look at this construction within the categorical setting of **Sets**, by first considering the following diagram:

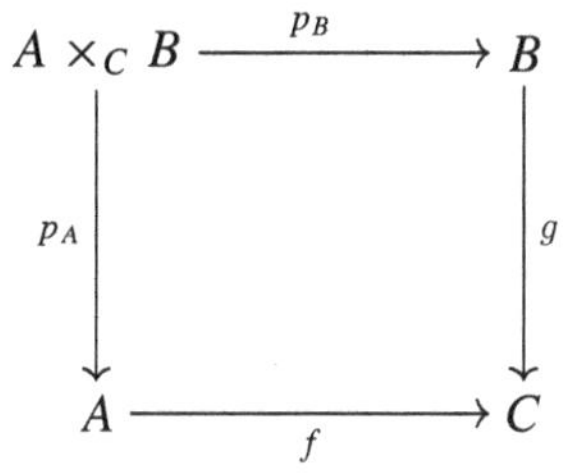

where p_A and p_B denote the projections over A and B, respectively.

As we learned in Appendix A, once a diagram is given in a category, a correspondent limit can be defined. In the case of the diagram above, taking the limit leads to the notion of *pullback* or *fibered product* of f and g. The reader should notice how the following definition refers only to objects and arrows in an otherwise undetermined category $\mathcal{C}$. In other words, the pullback or fibered product is here defined in a solely categorical manner (that generalizes the special notion of fibered product in set theory):

Definition 14 Let $f : A \longrightarrow C$, $g : B \longrightarrow C$ be a pair of morphisms in a category $\mathcal{C}$. A *pullback* (also called a *fibered product*) of f and g is an object $A \times_C B$ in $\mathcal{C}$ together with arrows $p_A : A \times_C B \longrightarrow A$ and $p_B : A \times_C B \longrightarrow B$, called projections, such that $f \circ p_A = g \circ p_B$, and for any object D in $\mathcal{C}$ and morphisms $h : D \longrightarrow A$ and $k : D \longrightarrow B$ such that $f \circ h = g \circ k$, there exists a unique morphism $< h, k >: D \longrightarrow A \times_C B$ such that the diagram

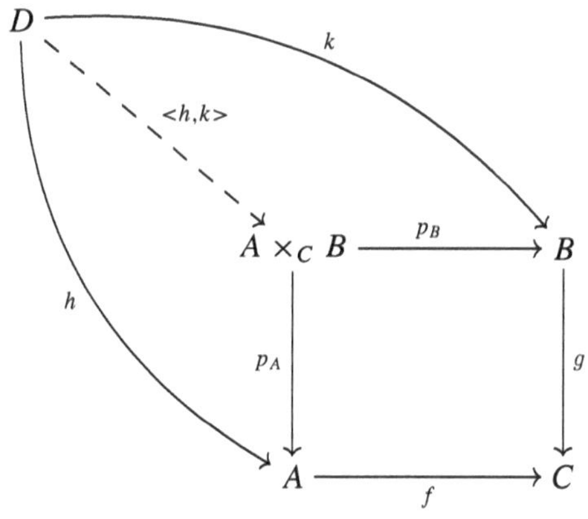

commutes.

We say that a category $\mathcal{C}$ has pullbacks iff every diagram

$$A \xrightarrow{\quad f \quad} C \xleftarrow{\quad g \quad} B$$

has a pullback.

One fundamental instance of pullbacks is given when g is a monic arrow, which, within a set-theoretical context, corresponds to declaring that—up to isomorphism of sets—$B \subseteq C$ (in the diagram above).[6] In this case, the pullback will return simply the pre-image of f as restricted to B:

$$A \times_C B \cong f^{-1}(C_{|B})$$

When this is generalized to categories, such a pullback along any monic arrow retains this feature of, roughly, the arrow as restricted to or fibered over the "subobject" of C designated by the monic (see below for the formal characterization of subobjects).

B.1.3 Exponentials

Given two sets A and C, we can consider the collection C^A of all the functions from A to C:

$$C^A := \{f \mid f : A \longrightarrow C\}$$

and define the *evaluation* function

$$ev : A \times C^A \longrightarrow C$$

such that

$$ev(x, f) = f(x)$$

This property of **Sets** can be generalized in strictly categorical terms as follows [5].

Definition 15 An object A of a category $\mathcal{C}$ is called *exponentiable* iff for every object X of $\mathcal{C}$ there exists an object X^A, called an *exponential*, and a morphism $ev : A \times X^A \longrightarrow X$ called an *evaluation*, such that for any $f : A \times B \rightarrow X$ there

[6]Throughout the present discussion we gloss over the distinction between subset and injective function (that is, subset equivalence up to isomorphism) for ease of exposition.

exists a unique $f^\flat : B \longrightarrow X^A$, called the *exponential transpose*, for which the diagram:

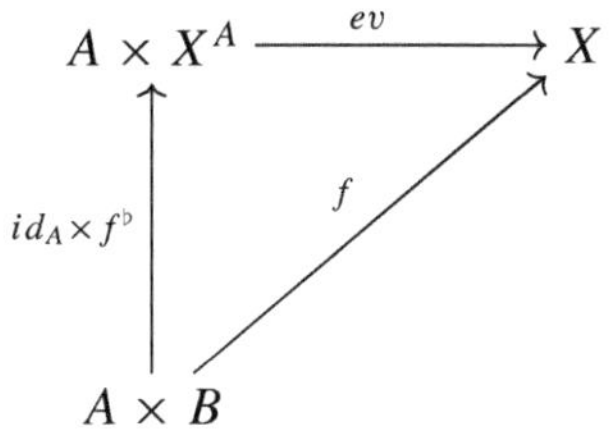

commutes. We say that category $\mathcal{C}$ has exponentials iff every object is exponentiable.

We leave as an exercise to the reader to show that the category **Sets** has exponentials.

Definition 16 A category $\mathcal{C}$ is called *cartesian closed* if and only if all objects of $\mathcal{C}$ are exponentiable (equivalently, $\mathcal{C}$ has exponentials) and $\mathcal{C}$ has finite products.

B.1.4 Subobjects

We can represent three nested sets A, B and C such that $C \subseteq B \subseteq A$ via the set-theoretical diagram

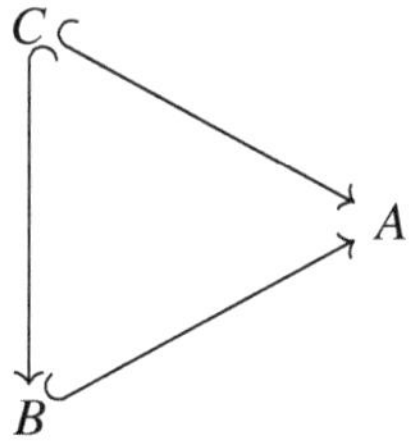

where the hooked arrow represents inclusion (an injection).

If we want to express the fact that C is (set-theoretically) equivalent to B, we need to replace the inclusion with a bijection, so that we get the following commutative diagram

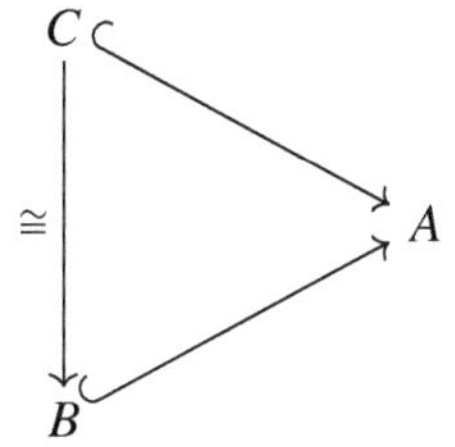

Let us now lift this construction into a purely categorical framework.

Definition 17 Two monic arrows x and y in a category $\mathcal{C}$ which satisfy

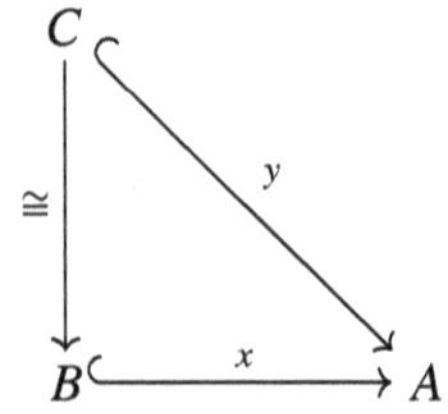

are called *equivalent*, which is denoted as $x \sim y$. The equivalence class of x is denoted as $[x] = \{y \mid y \sim x\}$. A *subobject* of any object A in $\mathcal{C}$ is defined as an equivalence class of monic arrows into A. The class of subobjects of an object A is denoted as

$$Sub(A) := \{[f] \mid cod(f) = A \; and \; f \; is \; monic\}$$

B.1.5 Subobject Classifier

Armed with the notions of monic arrow, subobject and pullback, we can now derive the categorical analogue to the set-theoretical notion of *characteristic function* for sets. Recall that, given a set X and a subset $A \subseteq X$, the characteristic function $\chi_A : X \longrightarrow \{0, 1\}$ takes every element of A into 1 and every element in the complement of A into 0. In other words, the notion of B being a subset of A can be expressed, in a set-theoretical context, using the characteristic function $\chi_B : A \rightarrow \{0, 1\}$, which is defined as follows:

$$\chi_B(x) = \begin{cases} 1 & if \; x \in B \\ 0 & if \; x \notin B \end{cases}$$

Working in the category **Sets**, the same idea is expressed by treating the inclusion $1 \subseteq 2$ as a stipulated monic arrow from an arbitary one-element set into an arbitary two-element set. This "true" arrow $\top$ then works to characterize monic arrows (set-theoretic inclusions) in exactly in the same way. To show this, let us consider a set **2** (that is, any two-element set), a terminal object **1** (that is, any singleton set), an inclusion arrow (a monic) $f : B \longrightarrow A$ that represents the subset B as an injective function, and finally a function $g : A \longrightarrow \mathbf{2}$. Because **1** is a terminal object, by definition there is exactly one arrow from B into it. We thus have the following data, diagrammatically arranged:

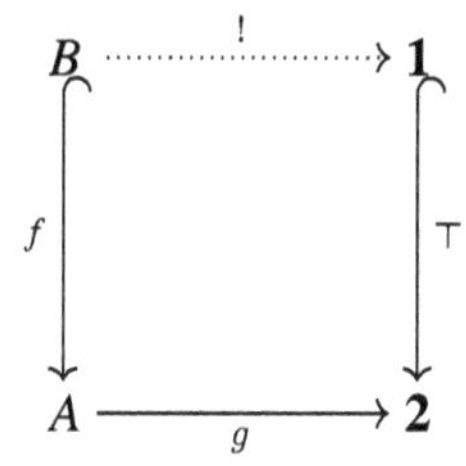

Then the reader should see that only if that diagram is a pullback is it the case that g in fact properly represents the inclusion (monic) arrow from B to A, that is, $g = \chi_f$:

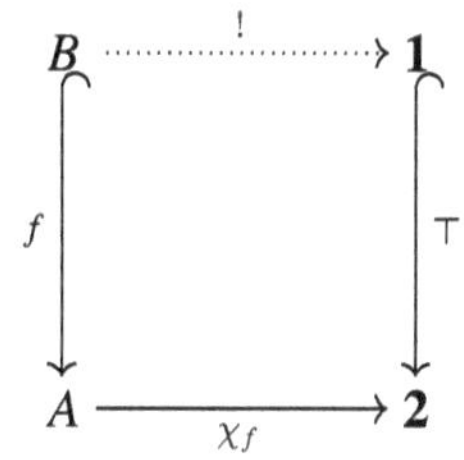

Notice the shift in notation from χ_B to χ_f. Thus the characteristic function in a categorical framework represents a monic arrow (a function), not strictly speaking an object. From a categorical perspective, a subset is a type of arrow, not an object.

Notice the heuristic of all of this: for any set B included in A as represented by an inclusion arrow f, the two-element set **2** is such that f is uniquely represented by a certain "special function" (χ_f) into **2**.

The generalization of this construction from **Sets** to a general category is immediate:

Definition 18 A *subobject classifier* or a generalised truth-value object is an object Ω in $\mathcal{C}$, together with an arrow $\top : \mathbf{1} \rightarrow \Omega$, called the true arrow, such that for each monic arrow $m : B \hookrightarrow A$ there is a unique arrow $\chi_m : A \longrightarrow \Omega$, called the characteristic arrow of m (or of B), such that the diagram

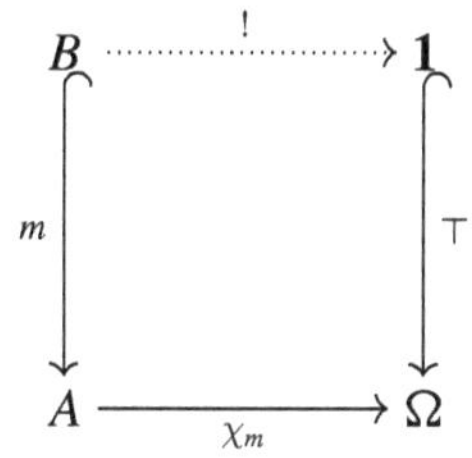

is a pullback.

Therefore, the generalization of the role of **2** in **Sets** to the role of Ω in a category $\mathcal{C}$ allows for a more differentiated structure of the analogue of set-theoretical complementation. In short, there may be a richer array of truth-values (more than just two).

B.2 Topoi

It goes without saying that the environment of sets and functions is a useful one for making a great variety of mathematical constructions. What if precisely the features that make this environment so useful for doing mathematics could be abstracted from those residual features of the set-theoretical universe(s) that may be at times unwieldy or even obstructive to mathematical investigation and production? In effect, this is what the abstraction from the category **Sets** to the more general notion of a topos accomplishes. We saw above how certain fundamental constructions in sets may be characterized in a purely relational or "arrows-only" fashion and then lifted to a more general categorical context. We are now prepared to characterize the result of this process through the categorical notion of a topos.

Definition 19 A category $\mathcal{C}$ is a *topos* (sometimes called an elementary topos) iff it is a cartesian closed category with a subobject classifier.

It is remarkable that such a simple definition can give rise to such a varied and powerful mathematical framework. We conclude by merely indicating several important examples of topoi:

- The canonical topos is the category **Sets** of sets and functions. In the previous sections we have indeed seen that **Sets** satisfies the elementary topos axioms, and this provides a helpful intuition of how the axioms together determine the possibility of certain basic mathematical constructions, even in topoi that are much more highly-structured or pathologically underdetermined from the standpoint of standard set theory.
- The category **Graphs** of directed graphs and graph homomorphisms is a useful environment for modeling networks, dynamical systems and many other related phenomena. **Graphs** is also a topos. The reader will find it instructive to show as an exercise that this category is indeed cartesian closed and to derive the necessary structure of its subobject classifier. As a helpful guiding thread, we note that the subobject classifier Ω of the topos **Graphs** is in fact the directed graph pictured here:

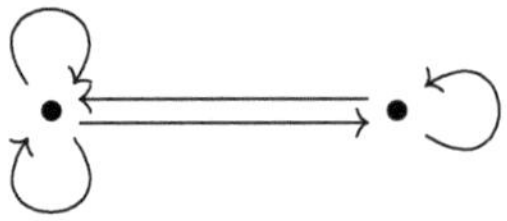

- An important class of topoi is given by the following construction. Take any small category $\mathcal{C}$ and then generate the category $\mathbf{Sets}^{\mathcal{C}^{op}}$ of contravariant functors from $\mathcal{C}$ into **Sets**. This category of functors (its objects are functors and its arrows are natural transformations) will always be a topos. As one example, take the simple category $\mathcal{D}$ consisting of just two objects A and B, the identity arrows Id_A and Id_B and two distinct arrows $f : A \longrightarrow B$ and $g : A \longrightarrow B$. Any contravariant functor F from $\mathcal{D}$ into **Sets** will correspond to a directed graph ($F(A)$ will be a set representing the graph's dots and $F(B)$ will be a set representing the graph's arrows; $F(f)$ and $F(g)$ determine *source* and *target* dots respectively for every arrow). A natural transformation from one such functor to another will correspond to a homomorphism of directed graphs. Thus the topos **Graphs** discussed immediately above may also be represented as a functor category $\mathbf{Sets}^{\mathcal{D}^{op}}$. In this way, the reader may perhaps begin to grasp how the basic conceptual approach of categories (characterizing the "internal" structure of objects through "external" relations to other objects of its type) carries over to the rich terrain of topoi as well.

References

1. R. Goldblatt, *Topoi: The Categorial Analysis of Logic* (Dover, New York, 2006)
2. A. Badiou, *Mathematics of the Transcendental* (Bloomsbury Academic, London, 2014)
3. S. Mac Lane, Moerdijk, *Sheaves in Geometry and Logic: A First Introduction to Topos Theory*. (Universitext) (Springer, Heidelberg, 1994)
4. P. Johnstone, *Sketches of an Elephant: A Topos Theory Compendium*. Oxford Logic Guides (Book 43) (2003)
5. R.P. Kostecki, *An Introduction to Topos Theory*. Preprint http://www.fuw.edu.pl/~kostecki/ittt.pdf

Bibliography

1. J. Adámek, H. Herrlich, G. Strecker, *Abstract and Concrete Categories. The Joy of Cats* (2004). http://www.iti.cs.tu-bs.de/
2. M. Barr, Ch. Wells, *Category Theory*. Lecture Notes for ESSLLI (1999). http://www.ling.ohio-state.edu/
3. G. Caterina, R. Gangle, Consequences of a diagrammatic representation of Paul Cohen's forcing technique based on C.S. Peirce's existential graphs. Stud. Comput. Intell. **314**, 419–443 (2010)
4. G. Caterina, R. Gangle, Iconicity and abduction: a categorical approach to creative hypothesis-formation in Peirce's existential graphs. Logic J. IGPL **21**, 1028–1043 (2013)
5. G. Caterina, R. Gangle, The sheet of indication: a diagrammatic semantics for Peirce's EG_{α}. Synthese **192**, 923–940 (2015)
6. B. Davey, H. Priestley, *Introduction to Lattices and Order* (Cambridge University Press, Cambridge (UK), 2002)
7. M. Heather, B. Rossiter, *Adjoint Exactness*. http://computing.unn.ac.uk/staff/cgnr1/anpa08.pdf
8. F.W. Lawvere, Functorial semantics of algebraic theories. PNAS **50**, 869–873 (1963)
9. E. Palmgren, *Semantics of Intuitionistic Propositional Logic*. Lecture Notes for Applied Logic. http://www2.math.uu.se/~palmgren/tillog/heyting3.pdf
10. C.S. Peirce, How to make our ideas clear, in C. Hartshorne et al. (ed.) *Collected Papers of Charles S. Peirce*, vol. 5 (The Belknap Press of Harvard University Press, Cambridge, 1960)

11. S. Philips, W.H. Wilson, Categorial compositionality: a category theory explanation for the systematicity of human cognition. PLOS Comput. Biol. **6**(7), 1–14 (2010)
12. A. Pietarinen, Peirce and the Logic of Image. Semiotica **192**, 251–261 (2012)
13. A. Pietarinen, F. Bellucci, What is so special about logical diagrams? Erkenntnis, forthcoming (2015)
14. A. Pietarinen, F. Bellucci, Existential graphs as an instrument of logical analysis. Part 1: alpha. Rev. Symb. Logic, forthcoming (2015)
15. P. Smith, *The Galois Connection between Syntax and Semantics*. http://www.logicmatters.net/resources/pdfs/Galois.pdf
16. K. Smith, *Adjoint Functors in Algebra, Topology and Mathematical Logic*. http://people.cs.uct.ac.za/~ksmith/adjoint.pdf
17. D. Spivak, *Category Theory for the Sciences* (MIT Press, Cambridge, 2014)